D0464800

SOLAR CATACLYSM

SOLAR CATACLYSM

How the Sun Shaped
the Past and
What We Can Do to
Save Our Future

LAWRENCE E. JOSEPH

HarperOne
An Imprint of HarperCollins*Publishers*

HarperOne

HarperCollins books may be purchased for educational, business, or sales promotional use. For information please e-mail the Special Markets Department at SPsales@harpercollins.com.

HarperCollins website: http://www.harpercollins.com

HarperCollins®, ✦®, and HarperOne™ are
trademarks of HarperCollins Publishers.

FIRST EDITION
Designed by Level C

Library of Congress Cataloging-in-Publication Data
Joseph, Lawrence E.
Solar cataclysm : how the sun shaped the past and what we can do to save our future
/ Lawrence E. Joseph. — 1st ed.
p. cm.
ISBN 978–0–06–206192–8
1. Solar activity. 2. Climatic changes—Effect of solar activity on.
3. Solar magnetic fields. 4. Sun. I. Title.
QB524.J67   2012
551.5'271—dc23   2012010158

12 13 14 15 16 RRD(H) 10 9 8 7 6 5 4 3 2 1

To Phoebe,
the sunshine of my life,
and Milo,
the star of my solar system

CONTENTS

FUTURE
177

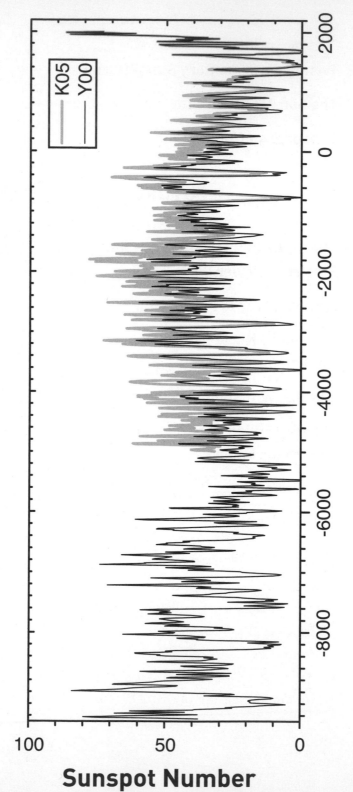

SOLAR ACTIVITY SINCE THE END OF THE LAST ICE AGE

Introduction

Did you ever find yourself staring at something for no good reason? Can't take your eyes off it even though you really have no idea why? Nerd that I am, I recently became fixated on a graph of the Sun's activity over the past 12,000 years or so, from the end of the most recent Ice Age on up through present day. The graph is from "A History of Solar Activity over Millennia," by Ilya G. Usoskin of the Sodankylä Geophysical Observatory in Oulu, Finland.[1] As you can see from the facing page, it's not much to look at, no fetching fractals, no graceful curves, just the quirks and jerks of some lifeless supermolten fusion behemoth 93 million miles away. One hundred and twenty centuries of solar boom and bust compressed into a single piece of paper turned sideways. After way too many hours of blinking uncomprehendingly, I taped the graph to the front of my computer printer, right next to the yellow Post-it note that says, "My Dad is the best Dad in the univers!!!!!"—a gift from my then-seven-year-old daughter, whose spelling, I must admit, was slow to embrace the silent *e*.

Phoebe's little love note cast the solar activity graph in a whole new light. After confirming that Usoskin's diagram is not some one-off outlier, but in fact genuinely reflective of other respected work in the field, I came to realize that the Sun's history is, after all, *our* history as well. For all intents and purposes, it is our only power source, our life-giving energy supply. Nearly every aspect of human existence is susceptible to changes in the Sun—in the long term due to climate change, in the short, utterly abrupt term due to the possibility of solar blasts causing massive blackouts, plus many other events, disruptive and benign.

My Moody Sun Hypothesis holds that nothing, including sunshine, is exempt from the laws of change. Random fluctuations in the Sun's behavior shape the course of history, affect our daily life, and determine our future in both subtle and cataclysmic ways. Our star is surprisingly changeable, moodier, in a sense, than most of us, scientists and nonscientists alike, had ever assumed. Each time the Sun erupts, goes dormant, or mixes up the spectral composition of its radiation, our environment changes, and, in one way or another, so do we. However, it is crucial to note that the Moody Sun Hypothesis is not of the deterministic ilk. Our fortunes, collectively and individually, are not anchored to the Sun's ups and downs. We simply respond to changes in its behavior with our ingenuity and survival instincts. The Sun is always a factor but rarely our master. We are not its slaves. (For more on the Moody Sun Hypothesis, see the Special Note that follows this introduction.)

Just how intimately our destiny is intertwined with that of our sustaining star was brought home to me while I was sitting in a sunny little park at the foot of Broadway on the southern tip of Manhattan, a few dozen yards from where General George Washington once surveyed New York Harbor to protect it from the British. Neither Washington nor I nor anyone else would ever have been nearly so interested in that particular vista had it not been for a bizarre quirk in the Sun's behavior at the end of the most recent Ice Age. Look back to the very beginning of Usoskin's graph, starting around 12,000 years ago. That's when temperatures soared due to a wicked spike in solar activity. Picture the moment some 12 millennia ago when the Hudson River, frozen solid for thousands of years, finally gave way. A trillion tons of ice melted into a mighty flood that inundated the isthmus known as the Verrazano Narrows and carved out the canals that now connect New York Harbor to the Atlantic Ocean. Had the Sun never melted the most recent Ice Age, New York Harbor might still be a freshwater lake, and would definitely never have become the grand ocean gateway for immigrants and commerce that made the city one of the greatest in the world.

Second in magnitude only to the giant post–Ice Age leap in the Sun's activity is the solar surge that began in the middle of the 19th century and that continues today. What if solar activity had instead troughed in the mid-1800s, and temperatures fell, not rose? New York Harbor just might have frozen up again, meaning that New York City would be landlocked for much of the year; no Big Apple, just a minor municipality. No Verrazano Bridge from Brooklyn to Staten Island for John Travolta and his pals to clown around on, and fall to death from, in *Saturday Night Fever*. Maybe the movie would never have been made at all.

The Sun also shares credit for the fact that the best-ever Brooklyn film was shot in English, not Norse, the language of the Vikings. During the 10th, 11th, and 12th centuries, the first European immigrants to what is now called North America did succeed in establishing settlements during the Medieval Warm Period. "The heyday of the Norse, which lasted roughly from AD 800 to about 1200, was not only a byproduct of such social factors as technology, overpopulation, and opportunism. Their great conquests and explorations took place during a period of unusually mild and stable weather—some of the warmest four centuries in the previous 8,000 years," writes climate anthropologist Brian Fagan in *The Little Ice Age*.[2] But then solar activity and terrestrial temperatures plummeted during the Little Ice Age, roughly 1300 to 1750. This drop had a profoundly chilling impact on the settling of the New World. The failure of the Norsemen to establish permanent colonies on our continent is less the result of any military, sociological, or cultural lapse than of the fact that formerly passable reaches of the northern North Atlantic became choked with ice for much of the year, thus preventing the seafarers from crossing the seas readily. "Vinland," as the Vikings had dubbed their Newfoundland colony, grew too cold to cultivate grapes, from which they made wine (*vin*). Without profits to be made from exporting wine back to Scandinavia, the dangerous trip was no longer worthwhile. As temperatures dropped during the Little Ice Age, so therefore did seafaring latitudes, favoring balmier climes. Colonizing power shifted southward to the British Isles, Portugal, and Columbus's Spain, essentially enabling those societies to reproduce themselves in the New World, unlike the barren intercourse of their Scandinavian forerunners. Thus it was George Washington, of English heritage, and not some descendant of Leif Eriksson, who ultimately led our nation's war of independence from overseas control.

The Moody Sun Hypothesis applies to the future as well as the past. The exploration and colonization of new worlds is just as susceptible to solar output variations today as it was a millennium ago. Astronauts are particularly vulnerable to radiation storms issued by the Sun, much as the ancient mariners were exposed to storms at sea. Both respond in pretty much the same way: hunker down and wait out the storm. But while the Vikings were frozen out of their destiny by the Little Ice Age dip in solar activity, a lull in solar activity would suit space travelers just fine. The fewer the sunspots, and the less powerful they are, the better the prospects for exploring and settling the Moon and Mars; relative space-weather calm is required for establishing dependable lifeline links between Earth and space colonies. A spike in sunspots can scramble long-range plans terribly, as happened in 1977, when Skylab 4, a precursor of today's International Space Station, was disabled when abnormally high solar activity caused Earth's atmosphere to heat up and expand. This increased the frictional drag on the satellite, causing its systems to malfunction. Fortunately, no astronauts were aboard at the time. The damage was irreparable, and Skylab 4 eventually fell out of the sky in 1979. This premature demise dealt a blow to NASA, which had expected the little space station to be the first port of call for the Space Shuttle program, when it was launched in the 1980s.

As explored further on, an emerging consensus of space-weather scientists argues that after the next solar climax recedes in late 2013, a long-term downshift in solar activity will commence. A pity, perhaps a national tragedy, that NASA's funding for deep space exploration is being slashed just as the Sun appears to be slipping into a hospitable lull. It brings to mind Jared Diamond's parable about Imperial China. China had been a great seafaring nation in the mid-1400s, but less than a century later, after a political squabble caused the Chinese to scrap their navies and shipyards, they fell behind the West in terms of political, economic, and cultural power for the next 500 years.[3]

LIGHTS OUT

Sometimes the Sun gets in a really foul mood and belches billion-ton blasts of plasma—essentially, supercharged energy in gaseous form—into the solar system. Most of these blasts go harmlessly into space, but a few, generally those launched from the northwest quadrant of the Sun's face, hit Earth.

Sooner or later, quite possibly within the next decade or so, we are going to get hit by a whopper that knocks out our electrical power grid and leaves up to 100 million Americans and countless others around the world without electricity for months or years. This prediction is according to a major report, *Severe Space Weather Events: Understanding Societal and Economic Impacts*, issued in December 2008 by the National Academy of Sciences in collaboration with NASA. The National Academy of Sciences is the closest thing there is to a Supreme Court of scientific opinion in the United States and much of the rest of the world. Created by Abraham Lincoln during the height of the Civil War and by far the most prestigious science membership organization in the world today, it is definitely *not* a fringe group. The Academy's report on the space-weather threat finds that solar blasts the size of the ones that hit Earth in 1859, 1909, and 1921—all of which hit before electrical power grids existed—would today cause the Mother of All Blackouts.

What sort of effects would we see? "Potable water distribution [would be] affected within several hours; perishable foods and medications lost in 12–24 hours; immediate or eventual loss of heating/air conditioning, sewage disposal, phone service, transportation, fuel resupply and so on. These outages would probably take months to fix, straining emergency services, banking and trade, and even command and control of the military and law enforcement," according to the National Academy of Sciences report.[4]

Most of us have been through a power blackout, though usually just for a few hours or days, not the marathon projected by the National Academy of Sciences. How dismal a world where outages had become the norm, where we, or at least those of us who survived, fell into the habit of singling out the "white times," when electricity was available, for special mention! That's just what will happen the next time a massive blob of solar plasma shorts out the electrical power grid, and therefore the societies it suckles.

People ask if experiencing massive blackouts would be like being thrown back into history, before society was electrified. No, it would be much worse than that, because those folks knew how to live without electricity. We don't. That's the paradox of progress: the farther we advance, the more dependent on our advances we become. Civilization can no longer exist without electricity. As an example, nuclear power plants, after a month or so of being deprived of normal electrical inputs from neighboring power stations, would lose their ability to cool their rods and wastes, and would blow up like a string of Chernobyls and Fukushimas, despoiling our air, water, and land. (More on that in chapter 13.)

STORY OF THE MOODY SUN

Our relationship with the Sun has had its ups and down. The earliest humans saw the Sun as a great sky god who ruled the heavens, the fierce, hot mate of Mother Earth. Then, as our conception of deity was humanized, the Sun came to be seen as God's most luminous physical manifestation, created by the Almighty to orbit Earth and to provide us with heat and light. Gradually it became evident that we orbit the Sun and not vice versa—all the more reason to suppose that the luminous creation around which God had ordained us to revolve was perfect, spotless, the purest product of heaven.

Go back and look again at the solar activity chart. This time use it as a Rorschach test. What do you see? An electrocardiogram? A jagged grin? I see the 12,000-year story of our relationship with the Sun. The main plot of the story is about how solar-driven climate change contributes to the rise and fall of civilizations. One subplot finds contemporary relevance in the utterly ancient practice of Sun worship, while another subplot grasps the implications of recent, massive demographic shifts to sunbelt regions around the world. There is even a folktale dimension, though we may not recognize it as such, which explains how sunrise and sunset, visually identical at a given stage, can make us feel so different. The plot twist at the end is how a single, random blast from the Sun may well destroy our way of life in the near future.

The story depicted in Usoskin's graph is backed up by plenty of facts. Since 1960, at least 18 spacecraft have successfully achieved solar orbit and have sent back valuable information after measuring, analyzing, and photographing the Sun's atmosphere, surface, winds, and storms. They have probed its spots and pinged its core. Many of these solar research spacecraft were sent out in conjunction with the International Heliophysical Year (IHY) 2007–2008, perhaps the most ambitious global collaboration in the history of science and truly the golden era of solar scholarship. IHY organized thousands of scientists in dozens of countries into scores of conferences and initiatives.

A veritable armada of solar research spacecraft have been launched in the past few years, relaying terabytes of data daily. In October 2006, for example, STEREO (Solar Terrestrial Relations Observatory) twin spacecraft were placed by NASA into orbit around the Sun in order to provide three-dimensional images with the goal of understanding how to predict solar blasts and other disruptive space-weather phenomena. In February 2010, the Solar Dynamics Observatory (SDO) was placed into geosynchronous orbit by NASA for the purpose of examining the effects that the solar storms identified by STEREO would have on Earth's atmosphere. SDO has provided the most spectacular and informative images of geomagnetic activity ever seen. In March 2010, HINODE, a solar research spacecraft developed and launched by the Japan Aerospace Exploration Agency, used its onboard telescope to examine the poles of the Sun for small but exceptionally powerful sunspot activity.

Why all the sudden interest?

Take a closer look at Usoskin's diagram. See how it spikes up from the middle of the 19th century through the present day? That's a storm. We are in the midst of one of the most massive, powerful, relentless solar storms in history, and there is no clear consensus on how much longer it will last. We do know that since the 1850s, the Sun has been pummeling our planet with nearly twice the average number of solar blasts per year. All scientific reconstructions of solar activity show a pronounced increase over the past century and a half, with some studies indicating that the modern era has been more tumultuous than any other period since runaway global warming reversed the most recent Ice Age a dozen millennia ago.

It is a basic law of physics, and common sense too, that energy injections excite a system, including the global ecosystem and the human civilization embedded within it. Since the Sun began acting up a century and a half ago, the world's population has more than quadrupled to 7.0 billion from the 1.5 billion plateau at which it had held more or less steady for half a millennium. Energy consumption has increased thirtyfold in that same century and a half. The Industrial Revolution, at least two world wars, the Space Age, the Information Age, massive migrations to sunbelt regions, global warming—up until now, all this upheaval has been attributed to human activity. But like the generations in the Middle East who have grown up knowing nothing but war, and Americans who came of age in financial boom times such as the Roaring Twenties and the Dot.com Nineties, our perception of life has been skewed by the tacit assumption that the extraordinary is ordinary, that the exceptional is the stable norm. In this book I'd like to suggest an additional explanation—that we have been hopped up on the solar equivalent of steroids.

The Moody Sun Hypothesis

My Moody Sun Hypothesis states that fluctuations in the Sun's behavior shape our history, daily life, and future in ways that most of us, scientists and laypersons alike, have never imagined. Normally, the solar cycle rises and falls over a period of about 11 years. Climaxes cause increased geomagnetic activity here on Earth, which in turn is associated with a variety of phenomena ranging from assaults on the electrical power grid and other technological infrastructures to increased threats of skin cancer to disruptions of our brain's ability to make logical decisions. Sometimes the 11-year rhythm is disturbed, and the Sun gets into a "mood." Troughs can last decades or even centuries, resulting in corresponding downturns in temperatures here on Earth. Even slight variations in the amount of sunlight and other forms of solar radiation can cumulatively impact Earth's climate, with outcomes ranging from Ice Ages to the warming period we are now in. Long-term booms and busts in solar output also influence birthrates, migration trends, and even trends in war and peace. This book is skewed heavily toward the consequences that changes in the Sun's behavior have for our planet Earth, particularly for the human beings living here today. But in fact the Moody Sun Hypothesis applies to the whole solar system, since that is the Sun's proper domain.

Fiction writers are usually asked about other writers who influenced them, but for some reason, writers of nonfiction rarely get the same question. The working assumptions seem to be that we are not artists and that, in nonfiction, prose quality is subordinate to factual content—that nonfiction is in essence a "service" genre, one in which substance is more important than style. Questions posed to nonfiction authors therefore tend to focus more on the topic at hand than on its presentation. Bruised artistic ego aside, I would like to note for the record that *Solar Cataclysm* started out in the early 1970s in the mold of *Catcher in the Rye* and *Portnoy's Complaint*. One day at the Danbury, Connecticut, state fair, Monica, my college girlfriend, chewed me out for being emotionally unavailable, unfair, uncommunicative, and a bunch of other *uns* that I can't exactly remember because I kind of spaced out while she was laying into me, which proved her point, I suppose. A pound of fudge was involved, and a really awesome blue-ribbon pig. That night I hit back with a killer affirmative defense: True, I wasn't any good at emotions because, hey, they aren't important. We think feelings come from deep inside us, but actually they just pass through us, continued my argument. Human beings are nothing more than sophisticated membranes constructed to absorb physical stimuli such as noise and heat and create corresponding biochemical end products such as anger, desire, and affection. No big deal. Understanding how *un*important emotions really are would help keep people from being tormented by them, I proposed, and in the long run would prevent fights like the (stupid) one she and I were having at the moment from ever happening again.

Never call your lover a membrane. That gaffe aside, our argument provoked some useful questions that would eventually, decades down the road, culminate in the Moody Sun Hypothesis. This hypothesis arose from the fact that I could not stop wondering how much of what we assume that we *create* is actually just *processed* by us, making us intermediaries rather than artists. How much credit can the radio take for the music it plays? Is culture, essentially the expression of emotion through artistry, also overrated? Back then I had no inkling that the Sun figured into the equation, just a vague sense that a lot of what we humans like to take most of the credit for really does not originate with us, or at least not to the extent we might think. Case in point is the aforementioned fall of the Viking empire—the collapse would not have happened had the Sun not cooled down to stymie the explorers' descendants by refreezing the seas.

The closest I came to connecting the Sun to any of this theorizing about human agency was through Linda Goodman's *Sun Signs*, a monster bestseller of the late 1960s that introduced astrology to American popular consciousness. The central tenet of astrology is that one's psychological and emotional makeup is largely predetermined by external forces—specifically, those that emanate from the heavens. On the face of it, this is darn near identical to what I had set out to prove. Sun-sign astrology is the most basic form, the kind on which most daily horoscopes in the newspaper are based. The Sun represents you, the conscious individual. Wherever it happened to be in the great 12-spoked wheel of constellations—Aries, Taurus, Gemini, Cancer, Leo, Virgo, Libra, Scorpio, Sagittarius, Capricorn, Aquarius, Pisces—at the moment you were born determines a lot about who you are, or so the thinking goes. And where the Sun happens to find itself on a given day, that's pretty much where your life is at. The other planets and the Moon contribute, but the Sun is the prime mover. That part I like: the Sun as star of the show—yours, mine, and everyone's. It is a complex and dynamic partner, beguiling, unpredictable, not just some immense radiator running on automatic.

But the whole astrology thing always seemed so oojie-boojie to me. It is a flagrant violation of the inverse square law, which states that the strength of the force between two objects varies inversely with the square of the distance between them. For example, the gravitational attraction between two objects 2 feet apart is only one-fourth the strength ($1/2^2 = 1/4$) that it would be if those same objects were 1 foot apart ($1/1^2 = 1$). If the two objects were 3 feet apart, the connection between them would be one-ninth the strength ($1/3^2 = 1/9$), and so on. So how could it be that stars trillions of miles away exert enough influence upon us to determine our fate? After all, $1/\text{trillions}^2$ = negligible. The refrigerator exerts a greater gravitational influence on you than the Big Dipper does. Many kids eventually make the same kind of calculation concerning Santa Claus: even if he takes only a minute per home, there are way more than a million homes in the world and nowhere near a million minutes in Christmas Eve, plus travel time.

Astrology, like Santa, does not add up, though there are times when the smart thing to do is to go along with it anyway. For the record, the Moody Sun Hypothesis has nothing to do with horoscopes or anything like that. Although I have been involuntarily impressed by several astrological "readings" done for me over the years, I do not understand or accept the precepts that underlie them. No celestial angle calculations, no star alignments, no squares, oppositions, or trines figure into this book. But whatever tradition they work in, give astrologers all the credit in the cosmos for realizing that our lives, culture, and history are indeed influenced, sometimes dramatically, by events in the heavens, starting with the Sun. Their theories about how this all might work do not conform to scientific principles, but there is no denying that these students of the heavens got there first, by thousands of years. In ancient times, astronomer-priests who could predict the movements of the heavens often ended up in positions of great power, presiding over immense observatories such as Machu Picchu in Peru and Angkor Wat in Cambodia. Their goal: predicting the Sun's behavior as regards its impact on us earthlings. Did they somehow unconsciously observe and/or intuit the basics of the Moody Sun Hypothesis?

Scientist and inventor Nikola Tesla, an oracle if there ever was one, felt it in his bones; Margaret Cheney writes in her biography of him:

With hundreds of thousands of volts of high-frequency currents surging through his body, he [Tesla] held in his hand this magnificent creation, a working model of the incandescent Sun. . . . The Sun, he reasoned, is an incandescent body carrying a high electrical charge and emitting showers of tiny particles, each of which is energized by its great velocity. But, not being enclosed in a glass, the Sun permits its rays to strike out into space. Tesla was convinced that all space was filled with these particles, constantly bombarding Earth or other matter.[1]

Tesla saw the Sun as the great particle accelerator in the sky, irradiating our lives in ways manifold and sundry—energizing us, threatening us, mutating us. He shocked his colleagues with outlandish estimates of the Sun's power, estimates that have since largely proven correct.

Contemporary solar physics lacks Tesla's passion, demystifying the Sun, deflating it right out of our psyches. The more information we gather about it, the less we seem to care. The Sun has been demoted, contextualized as just one of trillions of stars, unremarkable but for the random accident of its proximity to Earth. Although it is indisputably the star of the sky, the bold, blatant orb has lost its air of mystery. Nothing ever seems to threaten or disturb it—no more drama of the type that occurred in ancient China, where fear that an eclipse was the work of a dragon devouring the Sun prompted multitudes to make loud noises to frighten the dragon away. No more astrologers executed for failing to predict a solar eclipse, as happened to the hapless Hsi and Ho, or so the Chinese history-fable goes. No new Stonehenges erected to make the Sun roar back to life after fading away in winter. No more battles peacefully resolved by the shock-and-awe of a solar eclipse, as happened in ancient Greece, when the warriors of Lydia and Media looked up to the sky, saw day turn to night, and then dropped their weapons and huddled together for support. Nothing that we see in the Sun's movements shows any tendency for it to go anywhere or do anything odd. True, it does disappear at night, a fact that troubled some of the ancients mightily. But, like infants who gradually learn not to freak out every time their mother leaves the room, humankind got faith a long, long time ago that the Sun always returns.

OUR CHANGEABLE STAR

The Sun produces energy that is the rough equivalent of 100 billion tons of dynamite per second. It does this by a nuclear fusion process that transforms 700 million tons of hydrogen into 695 million tons of helium and a few other trace elements. The 5 million tons of lost mass is converted to energy, according to Einstein's immortal $E = mc^2$. The Sun radiates evenly in all directions, so only a tiny fraction of its output reaches Earth. But what is a trivial amount for our star is darn near everything for our planet, warming the ocean and the dry land, driving the water with currents and the air with wind. Sunlight, direct and reflected, creates a greenhouse effect by warming the carbon dioxide, methane, water vapor, and other gases in our planet's atmosphere. For all practical purposes, the Sun shines with unwavering intensity and will continue to do so until it burns out billions of years from now—that's been our working assumption. Likewise, fluctuations in the amount of sunshine received by Earth's surface are generally deemed to be caused by changes in season or by barriers in the sky such as clouds or fog, never by any variations in the primal source itself. But common sense says that the Sun moves through phases just as the rest of the universe does. All creatures, energy fields, and inanimate objects change all the time. To cease changing would of course be the biggest change of all.

To get why the Sun's output naturally fluctuates, stop thinking of it as a great solid disc shining down from the sky. Imagine instead a lamp revolving like those on top of police cars, yellow instead of red or blue. Just as Earth spins, so does the Sun. A spot on the solar equator moves at 4,400 miles per hour and takes about 25 days to complete a revolution; polar regions move much more slowly. The Sun's consistency is more gaseous/gelatinous than it is firm or solid. It's like a top with horizontal layers that somehow spin at differing rates. These differential rotations engender a dynamo effect that creates the Sun's magnetic field. One can readily grasp that this swirling, ultracomplicated, multi-tiered, multitextured mechanism radiates its energy steadily, though not perfectly so.

When we take the Sun for granted, we do so at our own peril. Solar physicists are just now learning that the Sun is far more turbulent and explosive than they had thought. To their shock and amazement, on August 1, 2010, an entire hemisphere of the Sun erupted. This is the equivalent of an earthquake simultaneously shaking North and South America, Europe, and western Africa. "Filaments of magnetism snapped and exploded, shock waves raced across the stellar surface, billion-ton clouds of hot gas billowed into space. Astronomers knew they had witnessed something big," writes Tony Phillips, a NASA space-weather scientist. Phillips explains that explosions on the Sun are not localized the way they are on Earth. Rather, they are chain reactions of solar flares, tsunamis, and coronal mass ejections. What is now known as the Great Eruption lasted 28 hours and traversed the face of the Sun. "All events were connected by a vast matrix of magnetic fault lines, where small changes in plasma flow can set off big storms, each setting the next one off like so many kernels of popcorn."[2]

Like any other star, the Sun is subject to its surroundings. Impacts from comets and meteorites inject minerals into the star, causing changes in its internal combustion processes. The chemical composition of the fuel it is burning at any given moment is an important issue, much the same way that the composition of foodstuffs is important to our own metabolism. Scientists classify the Sun as a Population I star, comparatively rich in heavy elements such as gold and uranium, as opposed to Population II stars, which are poor in elements heavier than hydrogen or helium. The more heavy elements a star contains, and the greater the variety of them, the more variable its burn, and therefore its resulting behavior, will be.

It gets harder and harder to believe that the "solar constant"—the amount of solar energy received at the very top of Earth's atmosphere—is so constant after all. It used to be an article of faith among astronomers that all wavelengths of solar radiation—radio, infrared visible light, ultraviolet, x-rays, and a bit of gamma—added up in power to 1,361 W/m^2 (watts per meter squared). This is pretty bright. Let's say your living room is 300 square feet, or about 30 square meters. If it received the same amount of energy as the solar constant, your room would be lit up by about 1,021 40-watt bulbs, 24 hours per day, 365 days per year. The trouble with the solar constant calculation is that the Sun is brighter at solar maximum than it is at solar minimum. Just how much brighter is the subject of passionate debate; measuring the variance with precision from Earth's surface is made next to impossible by clouds, atmospheric interference, and other uncontrollable factors. So NASA has taken to measuring solar irradiance from space, a methodological improvement but still far from perfect, since maxima and minima vary so unpredictably from cycle to cycle. Traditionally, astronomers assumed that the Sun fluctuates only about 0.1 percent over the course of the cycle. A 0.1 percent change in 1,361 W/m^2 equals 1.4 W/m^2. Add or subtract another 40-watt bulb from your living room's lighting scheme. This change would cause your utility bill to vary by about $25 per year. However, a growing number of solar physicists believe that the Sun fluctuates much more vigorously—approximately 0.5 percent over the course of each cycle—meaning another four bulbs to light up or turn off. Over time, these fluctuations add up significantly. Your utility bill would vary about $125 per year under this calculation. Now multiply that $125 by the billions of structures worldwide and you get an idea of how major a "minor" variation in the solar constant really can be.

Solar input to a particular region of the planet also fluctuates depending on how Earth is positioned in relation to the Sun. We tend to assume that the change of seasons has to do with how far we are from the Sun; we equate warm seasons with proximity to our star. In fact, it's the opposite: we are farthest from the Sun right around the Fourth of July, usually one of the hottest days in the year, and closest to it shortly after New Year's Day, one of the coldest. The seasons occur because Earth is tilted on its axis 23.5 degrees. When the Northern Hemisphere is tilted toward the Sun, it is summertime. The opposite inclination brings winter. (Little or no tilt occurs during spring and fall.) The Southern Hemisphere, of course, is on the opposite schedule.

WHEN SUNSPOTS POP

Eileen Ford, founder of the famous fashion modeling agency, would have grasped the idea of sunspots right away. Ford has preached that whatever you eat shows up on your face. The more Coke, chocolate, and greasy foods that are ingested, the more blemishes and skin conditions that erupt. In the Sun's case, it's not junk food, but junk magnetism—twisted and broken lines of magnetic force taken as if by a conveyor belt system from the Sun's surface down to its bowels and then back up again, all bloated and black—that causes sunspots.

Sunspots exist in the *photosphere*, the lowest level of the Sun's atmosphere. They are darker, cooler, and less radiant than their surroundings. One would think, therefore, that the presence of these tepid blotches would cause overall solar output to drop. Such is not the case, however. Sunspots serve as portals through which the Sun essentially belches immense globs of plasma. Whether or not the aggregate energy value of said globs is greater than the energy value of the luminosity lost as a result of the darkness and coolness of the spots is still open to debate, at least among those solar physicists interested in such things. What matters to the rest of us is not if the boiling oatmeal is hotter or cooler (net effect) when it bubbles up, but which of those bubbles will burst and spit in our eye. Similarly, scientists have a vital need to study sunspot patterns and trends, to inform their best guesses about which ones will explode, and how ferociously, in our direction.

We have no way of knowing what's the healthiest form of behavior for a thermonuclear behemoth that's 4.57 billion years old. You know how the spikes and plummets in Usoskin's graph look like one of those preprogrammed hill-climbing workouts on an exercise cycle? The Sun's fitness, like a cyclist's, may depend on traversing peaks and troughs rather than just pedaling along the flats all the time; there may be no single optimal level of activity. The sensible assumption is that it is thanks not only to our moody star's long periods of constancy but also to its ups and downs that we have managed to achieve the current stage of evolution, wondrous if for no other reason than it gave rise to you and me.

Recall that the solar activity graph portrays the leaps and plunges in the Sun's output over the past 12,000 years. How do Usoskin and his fellow solar physicists know so much about our star's tumultuous past? They count the planet-size dark blemishes that have routinely mottled the Sun's face from time immemorial. The more sunspots there are, the greater the solar output; that's the time-honored belief. However, counting sunspots turns out to be pretty complicated for two basic reasons. Sunspots often come in groups, making it difficult to know how many individual spots there actually are. The rule of thumb that astronomers use is that one group equals 10 individual sunspots. The second complicating factor is that the number of sunspots that can be seen (and therefore counted) varies with the strength of the observer's telescope. So in order to make the results historically consistent, solar physicists have to stick to those sunspots visible with simple telescopes and/or the naked eye.

Consistently recorded observations of sunspots extend back only to the early 1600s. Ascertaining the Sun's behavior all the way back to when the most recent Ice Age ended 12,000 years ago requires a whole extra set of forensic skills. Scientists have long known that solar radiation effectively blocks out cosmic rays—that is, energy from extraterrestrial sources other than the Sun—because rays from the nearby Sun are far more powerful than those that have traveled many light-years through the interstellar void. The more the Sun radiates, the fewer the cosmic rays that reach Earth. Cosmic rays are nonetheless distinctive, leaving a faint radioactive signature on all organic (carbon-based) substances they encounter. Using dating techniques based on radioactive isotopes (also called radioisotopes—radiation-emitting atomic variants of an element) such as carbon-14 and beryllium-10,

scientists examine ice core samples, tree rings, and other natural history artifacts for comparative concentrations of cosmic
ray–affected materials. The more carbon-14 and beryllium-10
at any given period, the more cosmic rays were getting through,
and therefore the weaker the Sun was at that time. Radioisotope
dating technology has become incredibly sophisticated—as precise, for example, as the forensics of DNA. Contemporary science
has brought us to the point where reliable assessments of radioisotopes can be made from samples only a few molecules in size.

Sunspots are not the only indicators of solar activity. Others
include faculae, filaments, prominences, flares, and radio waves.
Faculae are the color opposite of sunspots. These bright, white-
light patches, also occurring in the photosphere, are quite a bit
harder to see than sunspots, so we know much less about them.
They are not rigorously indexed, and therefore no hard historical information exists regarding their activity. The same lack of
formal statistical monitoring goes for solar filaments and prominences, spectacular million-mile "elbows" jutting into the interplanetary void. Charting such haphazard billowing historically
would be about as hard as trying to back-calculate the crackle
of a fire. Certain other solar phenomena are more quantifiable.
Flares shoot from the surface of the Sun on a daily basis. These
are monitored by a "flare index" that measures the total energy
emitted by these events. Radio-wave emissions from the Sun, also
monitored via a daily index, are considered a good measure of
general solar activity not directly related to sunspots. The "coronal index," compiled from monitoring stations all over the world,
provides a reliable assessment of total solar output. Other indices
compare the Sun to other stars, although such calculations are, of
course, complicated by the great distances between stars.

Fascinating as these ancillary solar phenomena might be, at the end of the day scientists have tacitly agreed to concentrate mostly on sunspots, which are more powerful, less transient, and thus easier to track and evaluate. Perhaps a more inclusive methodology for appraising solar activity will one day emerge. Any such refinement will likely heighten our awareness of solar change: the more powerful the magnifying glass, the more facets are revealed, and therefore the more opportunities there are for witnessing variations in those facets over time. There seems little doubt that future discoveries in solar research will serve to dispel any lingering allegiance to the shopworn belief in the Sun's aloof constancy, and instead bolster the more dynamic, engaged, and at times treacherous model of its role in our lives. The good news is that none of this variability poses a threat to the Sun's stability. Our own stability, however, is quite another question.

An old adage enjoins us to "make hay while the sun shines." So what hay can we make from the Moody Sun Hypothesis? Of what practical value is a supposition about the Sun? It will not prove of much solace when you take a loved one to the ER, although it may be quite helpful in reducing the need to make such trips in the future. It probably won't make you rich, but it could help keep you from becoming very, very poor. It might help explain why you are feeling so horny soaking up the vitamin D in the summer sunshine, but it won't offer much guidance in finding love.

Mostly, my Moody Sun Hypothesis is a new way of looking at things, meaning that the results are just starting to come in. "What a man sees depends both upon what he looks at and also upon what his previous visual-conceptual experiences have taught him to see," philosopher of science Thomas Kuhn reminds us.[3] Kuhn points out that making a transition to a new paradigm is an exciting process, particularly in the early stages, when one first renounces the old way of thinking. Many important discoveries come early on in the process, as though a pent-up demand for new insights were suddenly released. In this case, making the transition means breaking the habit of mind that assumes the Sun to be unchanging and irrelevant to our lives beyond the light and heat it normally provides so steadily. Of course, the hypothesis that the Sun's behavior influences our lives and even controls our destiny may look a bit radical and unprecedented today, but give it a generation or two and odds are that folks will be surprised that anyone ever believed otherwise.

> But, soft! what light through yonder window breaks?
> It is the east, and Juliet is the sun.
>
> William Shakespeare, *Romeo and Juliet*

Now is the time for the Sun to step out of the wings and into the starring role. With our growing understanding of the importance of our moody star's role in daily life, and of its myriad, often unpredictable influences, come a heightened desire and increased need to relate to it in more human terms. By being more aware of our attitudes toward the Sun, even those that are unconsciously held, perhaps we will find it easier to understand that we are in a relationship with our favorite star—a relationship that, like any other, needs work from time to time, especially right now.

PAST

Mountains of data exist about the Sun and the Earth, otherwise known as Sol and Gaia. What's in shockingly short supply is a sense of what all those facts mean for us personally. The first thing to know about the relationship between Sol and Gaia is that it cannot be understood in purely intellectual terms. Rather, theirs is a love story, an account of the sacred, profane, prolific bond between our planet and her sustaining star. We, the fair-haired children of that union, will never grasp entirely what the whole damn, wonderful thing is all about. Like Romeo and Juliet, Adam and Eve, Dante and Beatrice, Superman and Lois Lane, and Grandma and Grandpa, Sol and Gaia belong right up there with whatever other luminary couples might already inhabit your personal pantheon of lovers.

Could it be that the sacred human emotion called "love" is actually patterned after the supercharged interplay of Earth and Sun? Despite a penchant for drama (sometimes highly explosive like you wouldn't believe), our loving, quarrelsome parents have done the right thing and have stayed together through thick and thin for almost five billion years. The Moody Sun Hypothesis affords us an opportunity to see ourselves in a new light, as another kind of child. Of course you are the biological child of your parents. Perhaps you were raised by grandparents, adoptive parents, friends of the family who think of you as their child. Maybe you are inclined to extend "child" into the metaphorical realm, as in "child of the sixties." As we shall see in this section, we now have another identity, "child of Sol and Gaia." This is the literal truth.

1

Sol Has Loved Gaia for Almost Five Billion Years

Our little corner of the universe, the solar system—named after Sol, the Roman Sun god—coalesced 4.57 billion years ago out of a great cloud of molecular dust. Out of the confusion came the fusion-hot Sun, followed by more temperate objects such as planets, moons, and comets that spun off into the void, kind of the way that the main log in the fireplace keeps burning while hot bits are thrown from the fire to cool on the ground. (Asteroids came later, the fragments of shattered planets.) This dust-cloud theory of the solar system, more formally known as the Nebular Hypothesis, was first proposed in 1734 by Emanuel Swedenborg, the Swedish scientist–inventor–parliamentarian–theologian–

mystic. Although Swedenborg's hypothesis has been tweaked repeatedly over the past three centuries, it remains the scientific consensus on how Sol became king of the solar system, king of the sky, king of our physical reality. Sure, Sol is next to nothing within the context of the great Milky Way galaxy, and less than a speck of a speck of a speck of the universe. But in our tiny pond, he is the great golden sunfish.

The solar system's creation story parallels the opening narrative of the biblical book of Genesis, the 3,500-year-old work of genius that, with its seven-day parable of how God began with the void and ended up with the Garden of Eden, sagely foreshadows Darwin's premise that evolution progresses from barest simplicity to glorious complexity. In Genesis, God made Adam and then created Eve from Adam's rib. In the Nebular Hypothesis, the great dust cloud made the Sun and then made our planet, also known as Gaia, the primordial Greek goddess of Earth, from dust spun off by Sol. Adam and Eve mated and founded the human race. Sol and Gaia mated across space to create life as we know it, including the human race. Just as Adam and Eve begat Abel and Cain, so Sol and Gaia begat the two faces of nature—peaceful and murderous.

The relationship of Sol and Gaia vastly predates that of Adam and Eve and dwarfs it in cosmic importance, at least in my opinion. Yet for some reason our species, a newcomer on the evolutionary scene, has developed the habit of using itself as the standard of comparison for the much older, grander processes that gave us rise. Parents do not take after their children. Everything our species does, thinks, and feels cannot help but be patterned after the cosmic processes that created us. Do we human beings actually control our destiny? This question has been debated since the dawn of civilization. Emotions run high in this politically charged argument. God versus free will. Nature versus nurture. Society versus the individual. The ineluctable arc of history versus the creative chaos of the present moment. And now, Sol versus Gaia.

Sol is, well, moody at times, and his moods powerfully affect our own, along with the trends and shifts of our culture. Sol, of course, has no consciousness. Neither, most likely, does Gaia, tempted though one might be to attribute some form of collective awareness to her global pervasion of sentient organisms. Nevertheless, we find it useful to try to see things from our planet's perspective. From celebrating Earth Day to pondering planet-wide phenomena at the annual American Geophysical Union conclaves to convening 100-plus heads of state every 20 years at United Nations Earth Summits (in Stockholm 1972, in Rio 1992 and 2012), we have strived to master the global vantage point on everything from climate change to the thinning of the strato-spheric ozone layer to the planet's vulnerability to impacts, such as the one 62 million years ago that extinguished the dinosaurs. In so doing we have begun to empathize with our planet, asking ourselves not only what would Jesus do, but also what would Gaia do, think, feel. So it's high time we took into account Sol's role in all this. After all, he is footing the whole energy bill.

Fanciful as the notion might seem initially, the idea of an immortal, insensate bond between Sol and Gaia turns out to be not so unprecedented after all. Ancient Native American cosmology holding that Grandfather Sky and Grandmother Earth together begat human beings and all other living creatures is certainly more accurate and insightful than most modern cosmological metaphors, such as Buckminster Fuller's sterile and technocratic

"Spaceship Earth." But Sol and Gaia in love? Why impose human gender characteristics on inanimate objects such as planets and stars? Because the parallels are too obvious to ignore. Just as human love has a genuine electrochemical component—animal magnetism pulling, hearts racing, sparks flying, pulses pounding—so, conversely, does the bond between Sun and Earth have an aspect that transcends the merely physical. Mother Earth and Father Sun have for millennia been revered as deities in indigenous cultures around the world, and for good reason. Mother Earth teems with life, on land and in its waters, much as a female might bear and nurture her young. Father Sun's energy cocreates and then empowers that life, much as a responsible male might sire his young and then provide for his family. Regarding Sol and Gaia as a married couple with kids does not really anthropomorphize their relationship. Quite the contrary, their relationship has served, however unwittingly, as the template for the traditional man–woman human dynamic—although, truth be told, the cosmic duo hasn't always set the best example.

For one thing, Sol has not exactly been monogamous. In fact, he has kept a harem, nine planets rotating and revolving around him in the void. Only our planet, however, has given biological offspring to Sol. Gaia is unique in the solar system in her ability to convert solar inputs into complex living organisms, a fact that presumably earns her the status of number-one wife. Oh, Mars may have spawned a few bacteria way back when, and now they say that Jupiter's moon, Europa, may harbor some forms of life deep below the surface. Three cheers for mutant microbes. There's certainly nothing that remotely compares to the lush grandeur and staggering complexity of our global ecosystem. The other planets? Gaia is not jealous, and why should she be? They are barren of life, barren even of the ability to sense that they are barren.

Not that things were always easy for the young couple. The marriage of Sol and Gaia got off to a rocky start. For the first billion years or so after coalescing from the nebular dust cloud, the newlyweds had very little time for each other. Gaia convulsed with earthquakes and volcanoes, and with no atmosphere to protect her, all manner of space debris rained hell down from the heavens. Sol was faint and weak, far less powerful than he is today. He just wasn't in the mood. Fortunately for the young couple, Gaia was too molten to notice Sol's puny output; if anything, she would have been grateful for his inadequacy. The last thing the beleaguered bride needed from her husband during what is known as their Hadean (literally "hellish") era was more heat, given that she was convulsed with spasms and fever for the first chapter of her marriage.

Gradually Gaia's chaos subsided and an atmospheric cocoon of gas began to form. Steam condensed and pooled into oceans. Surface temperatures eased and then plunged. With Sol still struggling to boost his output, Gaia went from feverish to frigid. She would have frozen over completely were it not for the miracle known as the Faint Young Sun Paradox. Carl Sagan and fellow astronomer George Mullen calculated that during Earth's early history, the Sun was only about 70 percent as strong as it is currently.[1] They assert this based on what is known as the Standard Solar Model, which firmly holds that stars similar to the Sun brighten over the course of their lifespan, then peter out as they near death. A Sun as weak as the one envisioned by Sagan and Mullen would not have provided enough heat to keep Earth's oceans liquid—a big problem, because the presence of liquid water is absolutely essential to the formation of life as we know it. But somehow our fair planet did not freeze over. There is ample evidence of life dating as far back as 3.5 billion years, with the fossil record continuing on through the present day. That's the paradox: liquid water, and the life that it helped engender, should not have been able to exist on our planet, but it did, thank goodness.

Love stories have predictable arcs to them, as though they follow some set of natural laws. Do those same laws also apply to the celestial romance between Sol and Gaia? There are all sorts of explanations for how Gaia kept her water from freezing until Sol was finally able to deliver the heat. Sagan and Mullen attribute the bonus warmth to the felicitous effects of greenhouse gases, mostly carbon dioxide. As good fortune would have it, a particularly dense cocoon of these gases just happened to exist back then, keeping things extra-cozy. The more Sol warmed up, the more Gaia's greenhouse gas layer coincidentally thinned, keeping surface temperatures here on Earth from spiking. It was all very lucky. Suspiciously lucky.

Other explanations of the Faint Young Sun Paradox challenge the Standard Solar Model's central tenet that the Sun started out weak and grew stronger. Some point to meteorite data suggesting that the Sun was more active and luminous in its early days than the standard model would indicate. Meteorites that date back to the solar system's first billion and a half years do appear to have been struck with fewer cosmic rays (remember, those are rays from anywhere in space but the Sun) than younger meteorites. This observation is an important telltale because cosmic rays are blocked by solar wind, including the various forms of solar rays. The more active the Sun, the greater the solar wind and therefore the fewer the cosmic rays. Thus the conclusion that the young Sun was not so faint.

Some naysayers contend that the Sun started out larger than it is now and gradually shrank by exhaling a percentage of its mass by means of a vigorous solar wind. A powerful solar wind from the young Sun would have blocked out cosmic rays, widely believed to stimulate the formation of clouds, which, as we all know from personal experience, tend to reduce temperatures by blocking the Sun's rays. The net result would have been a reduction in our planet's cloud cover, thus warming things up, perhaps enough to keep Gaia's oceans from freezing over. The problem is that the meteorite/cosmic evidence is not nearly strong enough to support this explanation. Maybe the faint young Sun was not quite so faint as the Standard Solar Model would indicate, but that doesn't mean it was strong.

Regardless of the details, one can plainly see that the Moody Sun Hypothesis obtains here. Variations in the Sun's behavior from faint to robust determine the course of terrestrial evolution, not just of life-forms but also of our planet's surface and atmosphere. True, this all happened billions of years ago, but it would be absurd to assume that this primordial dynamic of the Sun–Earth relationship would just stop dead rather than continuing on through today and into the future. In fact, some scientists believe that the Sun gets 5 to 10 percent hotter per billion years, which is about how long it should take for the water to boil off our planet—although that could happen sooner if some catastrophe ripped away the atmosphere.

"Divine intervention" is not generally included in the group of hypotheses purporting to explain the Faint Young Sun Paradox. To me, this conundrum of how Sol and Gaia managed to work it out between them to keep things from freezing over is reminiscent of the miracle celebrated during Hanukkah, when a candle with a single day's supply of oil somehow burned for over a week. If evolution had had to wait an extra billion years until Sol got his act together and finally heated up, imagine how backward life would be today. Nematodes, at best.

Thank God for whatever it was that warmed Gaia's cockles. Perhaps the most imaginative suggestion as to what provided such miraculous assistance to our planet comes from Swedenborg, who believed that there are two Suns in the solar system, one that exists in the dreamworld and gives rise to all spiritual things, and another that exists in the material world and gives rise to all physical things. Because these worlds are so distinct, the two Suns have virtually nothing in common. Nonetheless, they do communicate, and are "conjoined by correspondences," as Swedenborg puts it.[2] To be open and receptive to these physical–spiritual correspondences in one's life is the path to enlightenment. However farfetched scientifically, the spirit–Sun notion does have its appeal. Not one but two heavenly sources of warmth and light consulting with each other to guide us when life's journey gets murky or, in Gaia's case, just too darn cold.

How did Gaia happen to disrobe at exactly the right pace? British atmospheric chemist James Lovelock doesn't buy that it's all just a coincidence. Lovelock invented the Gaia Hypothesis, a renowned scientific theory that Earth is more similar to a living body that adjusts and regulates its temperature and chemistry than it is like an inanimate rock or geological machine, as traditional Earth science holds.[3] He argues that early Earth's greenhouse gas bounty was no mere accident of geochemistry. Rather, the cozy cocoon was made by primitive microbes who had established a foothold on Gaia before she cooled down too far. These microbes proliferated wildly, as is their wont, collectively kicking out more and more carbon dioxide, methane, and other greenhouse gases. This kept the environment from cooling too precipitously, and therefore the water from freezing. Lynn Margulis, the renowned microbiologist who married Sagan but coauthored the Gaia Hypothesis, tends to agree with Lovelock, reminding us that 100 percent of living organisms emit gases, usually greenhouse gases, 100 percent of the time.[4] At some point Sol and Gaia got nice and cozy, and late into their second billion years together, husband managed to impregnate wife with his rays. The simplest bacteria—non-nucleated protoplasmic bits called prokaryotes— emerged and, as Margulis argues, immediately started pumping out greenhouse gas.[5] After another eon or so the prokaryotes merged with each other to form eukaryotes, microbes that had nuclei and that were capable of the more complex sets of tasks required to process the open sunshine. As Sol strengthened and Gaia's surface temperatures rose, new microbial strains emerged with the ability to metabolize greenhouse gases, thus thinning the cocoon and keeping temperatures from rising too rapidly. Cyanobacteria, the first organisms capable of photosynthesis, invented the color green to help them use the sunlight as a tool to create the simple nutrients their bodies demanded. In all, it took Sol and Gaia 3.5 billion years to raise the kids from mindless, centerless bacteria to complex plant and animal ecological communities and now to advanced technological civilization. Sol continues to provide everything Gaia needs, save for the errant interstellar gamma ray that arrives from another extraterrestrial source, and the steady sprinkling of ice and minerals that comes from comets and meteorites, which also originated with the nebular dust cloud. Gaia provides Sol with nothing in return, except, perhaps, awareness.

Question: Why is Gaia the only wife to bear Sol living offspring? Answer: Because she's not too close, not too far from her mate. What's known as the Goldilocks Hypothesis states that our planet is just the right distance from the Sun to create life. Not so close, and therefore not so hot, as to vaporize and outgas its water into space, and yet not so far away from the Sun's warmth as to freeze over and thus render its water unavailable for organisms to absorb. The power of Sol's rays plus the fecundity of Gaia's vital soils and fluids equals infinite new living entities, of which the human species is currently supreme.

You know when you move your hand slowly over your lover's body very close, tingling but not actually touching? Sol and Gaia are in constant contact like that. The two celestial lovers have not physically closed the distance between each other since they parted eons ago, though they seem to get a charge out of their special orbital connection. Every few months, every few decades—it varies—Sol tosses bouquets in the form of plasma fireballs that make Gaia's poles flush with beautiful auroras; sometimes the lights dance all the way down to her equator. (These are the same kind of fireballs that may one day soon decimate our civilization.) A glance at the solar activity chart at the beginning of this book reminds us that over the past century and a half the passion between Sol and Gaia has peaked, reaching climaxes unrivaled in thousands of years. Not bad for an old married couple. But check out the monster passions they released 10,000 to 12,000 years ago. That lovemaking was hot enough to melt an Ice Age.

2

Sunspots Melted the Last Ice Age

If you were chilly and sat by the fireplace and then warmed up, would you assume that you had warmed up all by yourself? No, you would probably grasp the simple fact that the fire had something to do with it. Yet historians, even brilliant ones, somehow fail to make that very simple connection when it comes to our planet.

"The Ice Age lasted for an unimaginably long time. Many tens of thousands of years. . . . But gradually *the Earth grew warmer* [italics mine] and the ice retreated to the high mountains, and people—who were by now most like us—learnt, with the warmth, to plant grasses and then grind the seeds to make a paste which they could bake in the fire, and this was bread," writes E. H. Gombrich in *A Little History of the World*.[1] Yes, temperatures on

Earth rose, but not of their own accord. They rose because the Sun got warmer. Duh!

No disrespect to Gombrich, a world-class scholar whose book *The Story of Art* is the landmark text in the field of art history. Ignoring the Sun's central, controlling role in our history is just something that historians seem to do. "With the end of the Pleistocene [Ice Age], about 10,000 years ago, the world's environments underwent many drastic changes. Thermal levels rose, the glaciers retreated, the seas rose rapidly, and many well-worn areas began to dry out" is how warming is explained in *The Columbia History of the World*.[2] The 1,200-page authoritative text found in library reference sections around the world contains not a single reference to the Sun or to its behavior! Same goes for the five-volume, 3,300-page reference masterwork *A History of Private Life*. In that voluminous compendium, one direct reference is made to the influence of the Sun on civilization.[3] (Apparently, sundials were quite popular in first-century-BCE Rome.)

"What's exciting is that we can now begin to link seemingly minor climatic shifts to all kinds of historical events in ways that were unimaginable even a generation ago. . . . [M]ost historians have tended to ignore climatic shifts, largely because as nonscientists they were unversed in the new climatological data," writes Brian Fagan.[4] With more than a dozen books on the relationship between climate and society, Fagan is unofficial dean of an emerging school of scholars focusing on the impact that climate changes have had on society. Fagan observes that the planet has always cycled between warming and cooling, a fact that is either ignored or negligently minimized in most current discussions of climate change. While the mechanisms of climatic undulation are fiercely debated, no one seriously denies the fact that the axiom "what goes up must come down" applies not only to gravity and momentum but also to the dynamics of earthly temperatures, both seasonally and over the very long term. A central question for this book is how climate cycles relate to solar cycles. Flip back to Usoskin's graph and note how it spikes abruptly right at the outset, maybe 12,000 years ago, and then again at 10,000 years ago. Those two spikes represent the one-two punch of heightened solar activity that halted the most recent Ice Age. First came the big thaw, then the big melt. Really, absent greenhouse gases and the like, what else but the Sun is powerful enough to have reversed an Ice Age that had kept much of the planet in a deep freeze for tens of thousands of years?

Over the eons, bitter cold Ice Ages have alternated with balmy times, some of which were distinctly hotter than our own today. This rise-and-fall cycling will likely continue until the end of earthly time. But since glacial and interglacial periods sometimes last hundreds of thousands of years, they are of limited value in understanding the climate changes that we have experienced over the past century and a half. We must therefore narrow our focus to the current interglacial epoch, known as the Holocene, the 120 or so centuries since the end of the most recent Ice Age. The Holocene may be considered a kiddy version of the planet's 4.57 billion–year roller-coaster past, a tiny reprise with less amplitude in its peaks and valleys of temperature and sunspots than in those of Earth's *entire* past. As noted, the Holocene's first and greatest peak of solar activity came some 12,000 years ago, with the end of the most recent Ice Age. Of course, those days were more dramatic than our own: temperatures jumped as much as one degree Celsius per decade, compared to the one or two degrees per century rate of increase we are experiencing today.

Ah, doesn't that Sun feel good? One tends to assume that the end of the most recent Ice Age came as warm relief, like winter thawing into spring. But Sol's amped-up energy blasts were anything but soothing for Gaia. Like any other major transition, the epoch was fraught with problems and possibilities. Although the retreat of glaciers took centuries, the flooding that occurred as these ice mountains melted was likely frequent and torrential. Storms unleashed by the extreme global warming must have been quite dangerous, much as they have recently become in our own climatically challenged era. But the effects were worse, because people back then were truly at the mercy of the elements. If not exactly cavemen during the last millennia of the Ice Age, most folks lived in primitive artificial structures, tents and such, which could hardly have been secure against megastorms—certainly not nearly to the extent that permanent housing is today. Particularly at risk were those who lived in shelters built on stilts in the middle of lakes, which served as moats. This was a common practice undertaken to defend against attacks by wild animals.

Profound ecological transformations swept much of the world as a result of that spike in solar activity:

With the diminution of the great Pleistocene [the Ice Age epoch from approximately 2.6 million years ago to 11,700 years ago] grasslands, the environments of the lower latitudes became more diversified. At the same time, many of the large game animals became extinct. Herds still roamed the remaining grasslands—zebra and antelope on the African savanna, bison on the North American plains,

guanaco on the Argentine pampas—but elsewhere the remaining species of large mammals became comparatively scarce.[5]

The population distribution not only of animals but also of human beings was deeply affected by Sol's Ice Age outbursts. The land and ice bridge over the Bering Strait that connected Siberia to Alaska melted and fell into the waves, thus putting an end to millennia of migration from Eurasia to the Americas.

The domino ramifications of those two powerful solar spikes 10,000 and 12,000 years ago are endless, example piled on top of example of how changes in the Sun's behavior control our history and shape our destiny. For example, the Moody Sun Hypothesis applies even to the prehistoric art world. During the last millennium of the Ice Age, cave painting flourished in Europe. "The most famous artworks are the cave paintings of France and Spain. Bison, mammoths, horses, deer, and other game animals are the most common subjects, but there are also a few human portrayals (including one dancer wearing a deer head mask) and some possibly symbolic geometric figures. . . . The cave paintings, engravings, and carvings of animals were apparently intended as hunting magic—'art for meat's sake,' as the great anthropologist E. A. Hooton put it."[6] But mysteriously, the solar outbursts seem to have put an end to cave art. Perhaps moderating temperatures inclined the artists to abandon their caverns and move outside, where, sadly for posterity, they painted on more perishable materials.

After banging and booming for the better part of 4,000 years, the Sun tuckered itself out. The first recorded trough in solar activity began around 8,000 years ago, and stayed low for approximately three millennia. The world changed dramatically when deprived of its usual solar input. For example, the Sahara went from being a vast humid stretch flourishing with fauna and flora to the desert it is now. This dry-out caused human and animal residents to migrate to the neighboring, more-cultivable region known as the Fertile Crescent, the crescent-shaped region in and around the Tigris and Euphrates Rivers that was largely contained in Mesopotamia, a land today known as Iraq. The reduction in solar output is believed to have resulted in lower levels of ambient growth, thus making foraging for food less productive, spurring the necessity of developing organized agriculture in what came to be known as the cradle of civilization. Humankind's transition from hunter-gatherer affiliations to a more structured agrarian society established the fundamental legal precept of private property, farmers' toil giving them ownership claim to the land they tended. The corresponding shift from itinerant to sedentary lifestyles affected the size and function of homes—those built to stay in one place were larger and more elaborate than their more portable predecessors. Humankind went from foraging to farming, from scouring the open terrain for predatory opportunities to committing to a specific piece of land as the means to survive, courtesy of the Sun. Agriculture then begat a great population boom, which in turn begat trade, which begat alphabets to document transactions and education to acquire the skills necessary to get the best deal.

What aspects of our early history *didn't* the Sun's behavior influence? The Moody Sun Hypothesis regards sunshine as a natural resource that, like any other resource, varies in both quality and quantity over the long term. While such fluctuations always affect the environment, and therefore the people and other life forms sustained by that environment, exactly how that all works out can be very hard to predict. As noted above, the weakening of the Sun 8,000 years ago led indirectly to the birth of agriculture, which ultimately proved to be an enormous boon for human-kind. Our species is unique in its ability to respond creatively to deficits in natural resources. Thus this hypothesis does not posit a straight-line correlation between human welfare and the strength or weakness of the Sun. Rather, we have adapted in response to each significant variation in solar output. Of course, until now we have had no idea that we have been partners in such an intricate coevolutionary dance for lo these many millennia. Our new-found knowledge will no doubt improve our performance.

3

The Long, Painful History of Sunspot Denial

Neither the sun nor death can be looked at with a steady eye.

François La Rochefoucauld

It hurts to look up at the Sun—always has, always will. Maybe the need to squint uncomfortably is one reason why we as a species have had such a devil of a time wrapping our minds around the fact that sunspots do, in fact, exist and that they might even be important. No problem with spots on the Moon. Call them craters, call them seas; they are pleasant to look up at and study—ah, the Sea of Tranquillity. But bring up the subject of sunspots and you are asking for trouble.

My guess is that the first time members of our species took serious note of sunspots was back 12,000 years or so ago, during the warming that ended the most recent Ice Age. As indicated by the sharp uptick at the beginning of Usoskin's graph, solar activity was extraordinarily high at that time, meaning that the Sun's face was covered with more spots than it is now, many of which were visible to the naked eye, just as sometimes is the case today. This change was all the more noticeable since, as far as we can tell, the preceding, frozen, low-solar millennia were characterized by a comparatively spotless Sun.

The earliest written mention of sunspots is on an ancient Babylonian tablet predating 1000 BCE that told of spots on the Sun appearing on the first of the month of Nisan, the Babylonian New Year, which began with the spring equinox, making Nisan the equivalent of March/April today. What that might have foretold for the future is anyone's guess, though it probably wasn't much, given that sunspots did not become a theme in subsequent carvings from that part of the world. In 364 BCE, Gan De and Shi Shen, a team of Chinese astronomers, blinded themselves while observing the Sun. Before their vision dimmed completely, they managed to describe dark blotches on the face of the Sun, interpreting them as sacred symbols from the heavens. Records of sunspots have since been kept, intermittently, by Chinese imperial court/government astronomers. These ancient court astronomers do not appear to have regarded sunspots as inherently good or ill. They just *were*.

The dogma of sunspot denial has its roots in Aristotle, who noted that the Sun, Moon, and stars never fall from the sky, nor do they move farther away. By contrast, the basic elements of our world—earth, air, fire, and water—are often in motion, either rising up or dropping to the ground. From this fundamental difference in behavior, Aristotle concluded that the heavens are made of entirely different stuff than Earth. Celestial objects are "ungenerated and indestructible and exempt from increase and alteration."[1] Aristotle also believed that the Sun is not innately hot, but rather gains its heat from the friction of its movement in orbit around us. One may excuse the venerable ancient's scientific errors, though not without noting that Aristotle's predecessor, Anaxagoras, correctly proclaimed that the Sun was an immense burning mass, and that his successor, Aristarchus of Samos, deduced that Earth orbited the Sun and not vice versa. Both had their reputations trashed by pro-Aristotle contemporaries.

Commonsense observations of the sky clearly suggest that the Sun rises and sets over, perhaps even revolves around, our planet. Aristotle drew this obvious conclusion and built his cosmology around it. But not unlike the psychological unfolding of an infant, who gradually learns that the world does not always revolve around him or her, the cozy vision of a perfect, Earth-centered cosmos fitfully gave way. Ptolemy, a Roman citizen of Egypt, noted that the Greek's Earth-in-the-middle worldview did not account well for the movements of the planets, particularly those outer planets that seem to stop and move backward, or in "retrograde." In *Almagest*, Ptolemy tried to solve this problem by declaring that the other planets did not exactly orbit Earth; rather, they snaked helix-shaped paths around a great circle of which our planet is the centerpoint.[2] At the center of this great orbital circle was not Earth, as Aristotle might have expected, but the "equant," a mathematical balance point that had no physical existence. Ptolemy's theory was complicated indeed, what with all the helixes snaking around imaginary centers, but he supported his hypothesis with a plethora of tables and graphs. One chart showed the Moon to be at various times twice as far from Earth as at other times, which means it should seem twice as large, which clearly is not the case. In fact, the full Moon appears to be exactly as large as the Sun, one of life's great coincidences being that the closeness of Earth's sole satellite should precisely compensate visually for how much smaller it is than our massive sustaining star. However intellectually flawed, Ptolemy's calculations proved quite useful in predicting certain planet appearances, eclipses, and the like, thus enshrining his work as the cosmological orthodoxy of ancient Rome.

Why the equant, and not the Sun, was for the next millennium considered the center of the cosmos is one of those eternal head-scratchers. Intellectuals have always been addicted to complexity, craving convoluted explanations like Ptolemy's when simple ones suffice. When Rome fell in the fifth century CE, the Aristotelian-Ptolemaic worldview was replaced by . . . nothing in particular, at least not until Arab translators revived some of the lost texts of Aristotle and Ptolemy in the late 12th century, helping stimulate the Renaissance.

In Europe, where science in those days was the province of the Christian Church, those ancient documents were adopted in part. Rather than getting bogged down with helixes and equants, the Church seized upon Aristotle's cosmology of heavenly perfection. Attention focused on the question of sunspots, which, the Church declared, could not and should not be. Consider the word of God:

And God said, "Let there be lights in the firmament of the heavens to separate the day from the night; and let them

be for signs and for seasons and for days and years, and let them be lights in the firmament of the heavens to give light upon the earth." And it was so. And God made the two great lights, the greater light to rule the day, and the lesser light to rule the night; he made the stars also. And God set them in the firmament of the heavens to give light upon the earth, to rule over the day and over the night, and to separate the light from the darkness. And God saw that it was good. And there was evening and there was morning, a fourth day. (Genesis 1:14–19, Revised Standard Version)

God made the Sun to rule over daytime, saw it was good, and that was that. Nothing in there about the Sun having pimples on its face.

Trying to deny the Vatican back in Renaissance days must have seemed every bit as foolhardy as trying to deny gravity:

From Cracow to Lisbon, from Edinburgh to Palermo, its [the Church's] teachings defined every man's faith. The spirituality of its saints shaped his devotions. Its priests baptized him when he entered the world and buried him when he died. Its authority and blessing touched rich and poor, priest and layman, noble and commoner, king and subject. The Church united Western European society in one common corps of Christendom, made it one body whose head was Christ.[3]

The monolith cracked in 1517, when Martin Luther, an obscure Augustinian friar, launched the Reformation by publishing

95 theses against corruption, blasphemy, and wickedness in what we now call the Roman Catholic Church. To many, Luther's theses must have seemed like a cross between "The Emperor's New Clothes" and a death wish. The fact that the heretic was not burned at the stake or zapped by a bolt from the blue demonstrated that the Church could be challenged head on. The haymaker punch to the enshrined worldview was thrown by Nicolaus Copernicus, who knocked Earth out of the center and replaced it with the Sun. The Polish astronomer had serious issues with Ptolemy, whose hyperconvoluted Equant Hypothesis made Copernicus squirm and turn to the Sun in desperation.

Copernicus listed assumptions he believed solved the problems of ancient astronomy. He stated that the earth is only the center of gravity and the center of the moon's orbit; that all the spheres encircle the sun, which is close to the center of the universe; that the universe is much larger than previously assumed, and the earth's distance to the sun is a small fraction of the size of the universe; that the apparent motion of the heavens and the sun is created by the motion of the earth; and that the apparent retrograde motion of the planets is created by the earth's motion.[4]

Substitute "solar system" for "universe" and his model works just fine today.

It was one thing to criticize the Vatican's human failings, as Luther did, but quite another to attack bedrock doctrine, or so Copernicus must have reasoned. Why else would the visionary have suppressed for 30 years his immortally convincing argument that

Earth revolved around the Sun and not vice versa? Copernicus first sketched out his Sun-centered notions in a pamphlet privately circulated sometime around 1514—not even a quarter century after Columbus's voyage experimentally proved that our planet was round, just like all those other spheres in the sky. (Odd that for centuries astronomers and philosophers had routinely assumed that the other planets and the stars were spheres, but not so Earth.) But it was not until several months before his death in 1543 that Copernicus published his boldest findings.

"In the middle of all sits the Sun enthroned," wrote Copernicus. "In this most beautiful temple could we place this luminary in any better position from which he can illuminate the whole at once? He is rightly called the Lamp, the Mind, the Ruler of the Universe; Hermes Trismegistus named him the Visible God, Sophocles' Electra calls him the All-seeing. So the Sun sits as upon a royal throne ruling his children the planets, which circle round him."[5] Like Luther, Copernicus gave voice to what many had thought privately, even suggested informally, but did not dare to publish—i.e., that the Sun, not Earth, was the cheese. Where would that leave all those earthlings supposedly made in God's image? Let's pray that Copernicus did not go to hell for his blasphemy.

By the beginning of the 17th century, Sol was a rising star, eclipsing Earth as the number-one celestial entity. For a brief while, the Sun reigned both immaculate and supreme. But as any good gossip columnist would have seen coming, stardom invariably brought out those who would shoot the star down. Instead of cameras, telescopes (which first appeared in 1608) were trained on Sol to reveal—what else?—sunspots, shaking Church doctrine to the core. That's odd, when you think about it, because some sunspots are visible to the naked eye, as evidenced by the multitude of ancient, pretelescopic observations. "Can it conceivably be an accident . . . that Western astronomers first saw change in the previously immutable heavens during the half-century after Copernicus's new paradigm was first proposed? . . . The very ease and rapidity with which astronomers saw new things . . . may make us wish to say that, after Copernicus, astronomers lived in a different world. In any case, their research proceeded as though that were the case," writes Thomas S. Kuhn in *The Structure of Scientific Revolutions.*[6]

True, the telescopes now showed more sunspots and in greater detail, vastly increasing the pile of evidence that the Vatican flacks had to deny. Beyond that, there must have been something coolly authoritative about the new optical instruments, devoid of any ideological agenda. If one day technology permits us to gaze upon the face of God, will we go through the same process of denial if imperfections are revealed? Let's hope that instead we have the humility to redefine perfection.

There is much kerfuffle over which Western scientist first discovered sunspots, "discovered" being a relative term, given that the ancients had gotten there 2,500 years earlier. Suffice it to say that, by 1611, about three years after telescopes came into use, claim to discovering sunspots had been laid by four men: Thomas Harriot, an English Anglican mathematician and explorer who made the first drawings and notes about sunspots but whose work was not published in a timely fashion; Johannes Fabricius, a Dutch Protestant astronomer who probably was the first to publish such findings, though with inadequate supporting documentation and analysis; Christoph Scheiner, a Bavarian Jesuit whose scientific work was superb but whose Christian orthodoxy and position in the Church hierarchy compelled him to deny what he saw, insisting that the spots were actually tiny planets; and Scheiner's archrival, Galileo Galilei, a Catholic who, though clearly not the first to discover sunspots, generally gets credit for having done so, with the publication of his *Letters on Sunspots*, written to his colleague Mark Welser beginning late in 1611. "Surely if anybody wants to imitate them [the sunspots] by means of earthly materials," Galileo said, "no better model could be found than to put some drops of coal on a red hot iron plate. From the black spot thus impressed on the iron, there will arise a black smoke that will disperse in strange and changing shapes."[7]

Why so much competition and jealousy surrounding the discovery of sunspots when simply declaring that they existed was taken as a middle finger to Aristotle's vision of heavenly perfection, and therefore to the all-powerful Church promulgating that doctrine? Perhaps it was some combination of love of science, belief that the truth about sunspots would come out sooner or later, and willingness to gain glory even at the expense of being punished. A goodly number of soldiers from the Holy See made the pilgrimage to Padua to meet with Galileo, a place in heaven no doubt reserved for that special someone who could make him recant his manifold heresies. In 1614, Jean Tarde, a French Jesuit canon, took his best shot, arguing that "the Sun is the father of light, and so how can it be diminished by spots? It is the seat of God. His house. His tabernacle. It is impious to attribute to God's house the filth, corruption, and blemishes of earth."[8]

Want to fit a square peg into a round hole? There is nothing like dogma to squeeze things in where they don't belong. Tarde proposed that what seems like changes in the size and shape of sunspots is actually the result of the convergence of planetoids—

small, planet-shaped objects—which, like grains of sand, sometimes clump together, other times fall apart. When Tarde named these tiny planets "Bourbon Stars," he had no idea that Louis XIV, of the house of Bourbon, would, as we shall see in the next chapter, chase them all away.

Galileo never backed down on sunspots. His exceptional brilliance and semiplausible claim to piety—all three of his children were born out of wedlock, but he was nonetheless a generous contributor to the Vatican—won him a good friend and protector in Pope Urban VII, who let the sunspot issue slide. Quite another matter, however, was the Paduan astronomer's headstrong defense of the Copernican view of a Sun-centered universe. Galileo's sin was in grasping the majesty of the Sun. "The Sun," he is widely attested to have said, "with all the planets revolving around it, and depending upon it, can still ripen a bunch of grapes as though it had nothing else in the universe to do."

Urban stuck by his friend until the latter wrote that Urban looked like a dunce for opposing Copernicus. In the now-famous trial, Galileo was convicted of heresy, and in 1633 he was sentenced to spend the remainder of his life, which would amount to eight years, under house arrest, with the additional penalty that none of his works was ever to be published again. But by then it was too late. The Copernicus-Galileo axis had forever elevated the Sun above Earth to the central position in the known universe, while at the same time besmirching its pristine character by verifying the existence of sunspots. This paradigm shift from Sun as perfect and secondary to flawed and central was too much for the Western mind-set to bear, akin in philosophical shock level to the finding of Big Bang theorists that in the beginning was not the Word, as John immortally wrote in his Gospel, but, rather, the Boom.

Before the Industrial Revolution, folks often looked to the sky for their inspiration and entertainment, much as we now look to the screen. Our ancestors had a complex love–hate, embrace-denial relationship with the heavens in general and the Sun in particular, attributing to it all sorts of spirituality, symbolism, and lore. Contemporary humans see the Sun as a giant ball of incredibly hot gas that maybe burps every now and then. Who among us today would consider pointing out a sunspot to be mean-spirited, even blasphemous, as though we were making fun of an embarrassing facial blemish, insulting to the Sun and therefore to God the creator? We denizens of the Postindustrial Age have few emotions or superstitions regarding the Sun. Sometimes it seems that we are too cool, calm, and contemporary about the whole thing for our own freaking good—except when it comes to the subject of the Medieval Warm Period (MWP), which lasted roughly from 900 to 1200 CE. Although a good thousand years have passed since the MWP, blood runs hotter than ever over the details of what transpired in the climate back then, and even over whether or not that warm period happened at all.

4

From the Medieval Warm Period to the Little Ice Age

Erik the Red, Leif Eriksson, and the boys must have drunk plenty of toasts to the Sun over the course of their lifetime. During the Medieval Warm Period (MWP), 900 to 1200 CE, Sunna, the Sun goddess, was in a very good mood. (Norse is one of the few mythologies in which the Moon is male and the Sun is female.) She smiled more frequently than she had in generations past, turning the air warmer, painting the landscape greener, and rendering the seas more navigable as temperatures rose. What better place to raise a mug to Sunna's radiance than in the Vikings' new (North American) colony, Vinland, where they planted, harvested, and fermented grapes into wine? But try toasting the

Medieval Warm Period the next time you are at a dinner party in that part of the world. If you are breaking bread with environmentalists concerned about global warming—and there are plenty of them in eco-conscious Scandinavia—your hearty compliments to their ancestors' balmy heyday may well provoke some icy stares.

The MWP is a very touchy subject in climatology circles. The mere fact that it happened at all is seen as a threat to the prevailing climate change orthodoxy that the emission of human-made greenhouse gases is almost exclusively responsible for our current bout of global warming. After all, back at the end of the first millennium, there were virtually no such emissions, except maybe for the smoke that curled out of wood, coal, and peat fires. But the last thing Al Gore and his fellow Nobel laureates at the Intergovernmental Panel on Climate Change (IPCC) want is for folks to start questioning how bad our current predicament could really be if our forebears survived, even prospered, under similar conditions.

How do we know how warm or cool it was a thousand years ago, anyway? That's a good 600 years before the thermometer was invented. To be sure, radioisotope dating techniques are quite useful in reconstructing ancient climate trends. Famed chemist Harold Urey discovered that ice core samples could be dated by the amount of radioactive oxygen (O_{18}) that was captured in air bubbles. The less there is of this rare but core component of elemental oxygen, the older the sample is. Same basic idea as carbon-14 dating. The good news is that scientists can make some pretty good guesses about ambient air temperatures by dating ice cores and then examining their characteristics, such as the amount and kind of dust they contain. The bad news is that such estimates are reliable only for the Arctic region from which the samples were obtained, usually Greenland.

Greenland is the Mecca of paleoclimatologists, as abnormally large in their imagination as it is on the Mercator projection that most of us tend to consult, the map on which the modest-size island is inflated almost to subcontinent status. Extrapolations from ice cores mined in Greenland underpin a far greater share of our beliefs about climate history than most researchers would care to admit. The island is close to the United States, so it is easy for American scientists to get there. The U.S. military is installed there and thus provides a rudimentary infrastructure and even support service. The locals don't seem to mind the extraction of bits of their native ice, so it all works out hunky-dory. Thus the powerful incentive to make the argument that, as Greenland went, so went the world. Wouldn't that be convenient?

Fortunately for those who wish to understand the MWP, the Viking heyday was centered in Iceland and Greenland, meaning that the ice core samples are probably very accurate reflections of what went on in that particular era and region. Scientists believe that there was a significant rise in temperatures in Greenland from the 10th to the 13th centuries, pretty much the span of the warming period in question.[1] Historical evidence, such as that collected from journals being kept at the time, indicates that Europe experienced a warming climate, including dry summers and mild winters, during the 11th through 14th centuries—not a perfect overlap, but close enough to be related. Question is, did the rest of the region, hemisphere, and/or world get warmer as well? Supporting data seem necessary for any area farther away than the northern North Atlantic region. One way to provide reliable results is to drill cores from the ice caps of mountains in equatorial or midlatitude regions—for example, the Andes range—and compare those findings with Arctic samples that cover the same range of dates. Alternative data might also be acquired by examining coral reefs, which grow a new band of calcium carbonate ($CaCO_3$) every year. The less the proportion of radioactive oxygen (O_{18}), the warmer the year was. The bad news is that these sophisticated transspatial, transtemporal, transmethodological comparisons involve a lot of guesswork and expense. Such investigations are often inconclusive.

What if the MWP never happened at all? Enter the "hockey stick graph," one of the most famous diagrams in the history of science, all the more notable for being statistically flawed. (You have probably seen this graph in the documentary film *An Inconvenient Truth*, or in other programs on climate change.) In the late 1990s, geoscientist Michael E. Mann, then of the University of Massachusetts, and colleagues set out to sort through and synthesize as many scientific reconstructions of global temperatures for the past millennium as they could, about 70 different data sets in all, using ice cores, tree rings, oxygen bubbles, coral reef bands, and just about any other temperature proxies they could find.[2] The idea behind this ambitious project was that if all these disparate measures were in basic agreement, then science would have a pretty firm grasp on what the global climate had done over the past thousand years. Even more important, this research would give a solid context for understanding what is happening in our climate today.

Mann and his colleagues ran all their data through a statistical software regimen called PCA (principal components analysis), which is supposed to extract the highlights of the data. It's kind of like a juicer, but for numbers. They found that there had been precious little mean temperature variation from the year 1000 to 1900. No indication of a Medieval Warm Period was found. But from 1900 to 2000, the period when greenhouse gases really started accumulating, temperatures rose sharply. The PCA program printed out these results in the form of a graph that, for the first 900 years, stays more or less flat, like the shaft of a hockey stick. After the year 1900 it takes a sharp diagonal jag, jutting up and to the right like a hockey stick's blade. Wow, awesome: the climate change Stanley Cup is what Mann and his team seem to have thought they won.

But wait a minute. What happened to the MWP? If there was no global warming back then, how did those Vikings sail around the northern North Atlantic? In ice-cutters? Climate change naysayers Willie Soon and Sallie Baliunas of the Harvard-Smithsonian Center for Astrophysics rushed to the MWP's defense. They quickly produced a study of surface temperatures over the past thousand years that was based on moisture content in the air. Water vapor levels are indeed an indicator of ambient temperature; the warmer the air is, the more moisture it will hold. Concomitantly, moisture is also an important means by which the Sun heats up the air, as the water droplets absorb the sunlight. However, although humidity and temperature are related, they are not perfectly correlated. Sometimes humidity actually buffers temperature extremes. Consider that the highest and lowest temperatures in the world have been recorded in arid locales, the Sahara and the Antarctic, respectively. Nevertheless, Soon and Baliunas stuck by their methodology, insisting that there are two blades on the thousand-year hockey stick, one at either end.

"We conclude that the available scientific data do not support the claim that the 20th century was the warmest or most unusual of the millennium," wrote Soon and Baliunas in *Lessons and Limits of Climate History*.[3] Really? And that sweeping abnegation of contemporary climate science and environmental policy is based on . . . water vapor? A more diplomatic phrasing would have been, "The twentieth century is one of the hottest climatic periods of the last millennium." Same factual essence but opposite rhetorical slant, though certainly not the confrontational tone that the (deeply right-wing) Marshall Institute and the American Petroleum Institute paid Soon and Baliunas good money to trumpet. Also unacceptable to the naysayer extremists would be the phrasing, "We have not experienced a warmer or more extreme climatic interlude than the twentieth century for at least 800 years." In other words, not since the end of the Medieval Warm Period.

Gore/IPCC saw red and swiftly demolished the Soon-Baliunas study, condemning its reliance on atmospheric water vapor as a temperature proxy and proving, at least to their own satisfaction, that the naysayer study could not be duplicated and was therefore invalid. The hockey stick went back to having one blade. Still, it was impossible to shake the feeling that something hinky was going down. Canadian scientists Stephen McIntyre and Ross McKitrick obtained the statistical software used by Mann and his team and then tested it by doing what is known as a Monte Carlo experiment, in which meaningless, random data is fed into the program. Out popped a graph in the shape of . . . a hockey stick! Apparently Mann and Company never tested the software with this standard procedure, or if they did, they ignored the results. Turns out that this particular type of statistical software thinks that as long as there's not *too* much fluctuation in values, as is the case with global mean temperatures (which generally vary less than a degree or two Celsius per millennium), data sets that otherwise bear no resemblance to each other all graph out to look like hockey sticks. So when McIntyre and McKitrick went back and corrected for the errors made by Mann's team, the Medieval Warming Period returned in all its jutting glory, temperature spikes and all.

How perfectly poetic that it took two Canadians, McIntyre and McKitrick, to break the phony hockey stick. And how perfectly awful that *Nature*, perhaps the world's leading science journal, refused to publish their findings. The referees found no fault with the scientists' methodology. They just did not think this discovery was important enough. Not important enough—the fact that the central, iconic graph of the global climate change movement, the movement upon which trillions of dollars of policy decisions were being made, might well be profoundly flawed? So McIntyre and McKitrick self-published their paper on the web, along with the referees' comments.[4]

Their paper languished in obscurity until Richard A. Muller, a grand old physicist from the University of California at Berkeley, swooped in. Writing in *Technology Review*, a publication of the Massachusetts Institute of Technology, Muller (who, incidentally, has been published in *Nature* many times) declared the Canadian scientists' discovery to be profoundly important:

If you are concerned about global warming (as I am) and think that human-created carbon dioxide may contribute (as I do) then you still should agree that we are much better off having broken the hockey stick. Misinforma-

tion can do real harm, because it distorts predictions. . . .
A phony hockey stick is more dangerous than a broken
one—if we know it is broken. It is our responsibility to look
at the data in an unbiased way, and draw whatever conclusions follow. When we discover a mistake, we admit it,
learn from it, and perhaps discover once again the value
of caution.[5]

So how many blades does the thousand-year hockey stick *really*
have? Two—the MWP and the present—when the graph is of
temperatures in the Northern Hemisphere, particularly North
America and Europe, including, of course, the Vikings' ancient
domain. As for the rest of the world, maybe a blade and a half.
Unlike the Little Ice Age (LIA), which was felt far and wide, proof
of the MWP as a global phenomenon is much harder to come by.
Could be simply that not enough research has been done and/
or that sufficient data from far-flung sources in the Southern and
Eastern Hemispheres from a thousand years ago are simply unavailable. Investigators interested in that period of history might
do well to look for regional pockets of warming comparable in
scope to those that happened in the northern portions of Europe
and North America. Until they make such discoveries, however,
solid evidence of the existence of the MWP is pretty much limited to the greater Viking domain.

Perhaps the Gore/IPCC contingent will ultimately prove correct in their contention that the MWP did not occur on a global
scale, and that whatever warming impact it had was so small
and localized as not to boost overall global temperature totals. A
glance back at Usoskin's graph confirms the lack of solar activity
spikes during the MWP, in the 10th through 13th centuries. Fur-

thermore, there's no denying the Western tendency to see world history as being what happened in Europe and North America. Yet there is also no doubt that the MWP happened regionally, and that it was far too crucial a turning point to be overlooked. In the meantime, erstwhile MWP deniers might take solace from the fact that in spite of all the balmy good times, the era was in its own way disastrous. Ice caps melted, tree lines climbed, insect vectors shot like arrows. Most significant, the turbulent North Sea rose by almost two and a half feet, and huge swaths of land in what had been (and are now once again) Britain, the Netherlands, Denmark, and Germany were inundated by the waves. Should comparable flooding occur today, millions of people would be dislocated and economies would drown.[6]

"A book should serve as an axe to smash the frozen sea within us" is one of Franz Kafka's most memorable quotations. Think of the MWP as an axe smashing the icebound North Atlantic and freeing the Vikings from frozen limbo to fulfill their destiny. Yet just because the seas had become passable doesn't mean that the Nordic navigators could tell where to go on cloudy days, of which there were still aplenty in what, geographically speaking, is basically suburban North Pole. When visibility was clear, the Sun, stars, and landmarks were useful to reckon from—but how, historians wondered, did the Vikings, sailing centuries before compasses came to be used on ships, keep their bearings when the sky was overcast?

There are two types of rock known as sunstone, both of which serve as crystalline portals that polarize light, pierce the clouds, and reconnect the user with the hidden Sun. The beaches of Norway are littered with pieces of one type: cordierite, a mineral that changes color from blue to light yellow, the color of sunshine, when pointed toward the Sun. The other type of sunstone, known as Iceland spar, is a rhomboid calcite crystal. Viking navigators would put a dot on top of the spar and then look through it upward at the sky until two dots appeared. The navigators would then rotate the sunstone until the two dots were exactly the same shade, at which angle the upward-facing surface of the stone would be pointing in the direction of the Sun. "How wonderful if it were [Icelandic spar]: the eyes of trilobytes—the earliest eyes we know of, five hundred fifty million years old—have lenses made of the same stuff," writes Simon Ings in *A Natural History of Seeing*.[7] Recent research shows that even at the approach of dusk, the Sun's direction could be calculated by sunstones to within a few degrees, thus enabling the intrepid sailors to get a rough fix on their position. But when it comes to thousand-mile voyages, a few degrees leeway is quite a substantial margin of error, especially in the minimally maneuverable Viking longships, which had neither keel nor aerodynamic sail. Having myself been associated in 1992 with a modern-day Viking ship mission from Norway to Brazil and then on to the Mediterranean, wherein one of the two ships was lost at sea (all hands rescued), I can tell you that any unplanned extensions of the journey were fraught with peril; historians estimate that as few as one in six such ships and their crews survived their voyage during the Viking era.

The ability of the navigator to use his sunstone portal to sense the Sun's location precisely might well have meant the difference between life and death for the crew. In "Iceland Spar," the title of the second section of the rambunctious novel *Against the Day*, Thomas Pynchon nimbly portrays this intimate, almost psychic bond—not an act of worship, exactly, but nonetheless a pure expression of faith by the navigator in his connection with the unseen Sun, an invisible guiding light. The difference is that Pynchon saw the crystal as a means by which human beings could voyage interdimensionally, while the Vikings no doubt channeled the Sun to help them remain safely within the earthly realm. Funny to think that the rough and tumble Vikings were more deeply into their crystals than any New Ager could ever be today.

The Vikings used their sunstones to bond with Sunna, singer of charms, the "bright bride" of heaven. One can only imagine the fervent seamen's fantasies about their Sun goddess as she led them across the ocean and, through the sunstone, showed herself to them even when she could not be seen in the sky. Who would be brave enough to save Sunna from Skoll, the angry wolf who daily pursued her from east to west in order to devour her? Norse mythology tells us that again and again the Sun goddess escaped, although the chase tired her out, causing her to grow faint and eventually fall to the hungry wolf. Thus, the MWP was followed by the Little Ice Age, which plunged the Vikings back into frigid obscurity. As Sunna's warmth and light shrank away, with it went the vitality of Nordic culture. Has she returned today as the hot breath of carbon dioxide? Will she be, *should* she be, embraced for her warmth, or is it humankind's duty to devour her as Skoll did almost a millennium ago?

THE LITTLE ICE AGE

"No one knows exactly what drives the climatic pendulum. Most likely, small changes in the earth's obliquity [angle of inclination] trigger climate changes. So do cycles of sunspot activity. For instance, a dearth of sunspots during the 17th century marked a period of markedly cooler climate during the height of the so-called Little Ice Age," writes Brian Fagan, the climate anthropologist.[8] In contrast to the MWP, which lacked noteworthy trends in sunspot activity, there was indeed a steep falloff in solar output to levels far below the norm during the Little Ice Age, 1300 to 1750. Usoskin's graph illustrates the depth of the troughs.

We know that the LIA was the last time glaciers advanced down mountains around the world. We also know that an enlarged polar ice cap sent frigid air throughout the Northern Hemisphere, weakening the jet streams and pushing them further south than they are today. Climatologists widely believe that the LIA was colder than our world now is, as well as being colder than was the Medieval Warm Period that preceded it. But how cold was it, exactly? And what was the windchill factor? It's hard to say: determining precise surface temperatures was impossible before thermometers were invented at the turn of the 17th century, about two-thirds of the way through the LIA.

The first thermometers came into use at about the same time as telescopes. Galileo introduced his "thermoscope," an ingenious if unwieldy precursor to the thermometer, in 1593, about 15 years before debuting his telescope. Other rudimentary thermometers were not very accurate, particularly the prototypes that used water and/or alcohol rather than mercury, which expands much more smoothly, thus allowing finer gradations of the temperature scale. Daniel Gabriel Fahrenheit, a German chemist, is credited with having made this liquid metal innovation in 1714. Whatever craziness it was that possessed Fahrenheit to set the freezing point of water at 32 degrees and the boiling point at 212 degrees seems fortunately to have eluded Anders Celsius, who in 1742 much more sensibly set the key values of freezing and boiling an even

100 degrees apart. Unlike today, when a new and useful invention can spread worldwide in a matter of a few years, mercury thermometers, considered a pressing necessity by only a comparative few, did not really go global until the 19th century. That means we have only 200-odd years of reasonably reliable, sufficiently far-flung temperature readings on which to base our most basic assumptions about the history that led to our climate today.

Most of what we know about the climate of the LIA comes from historical data such as notations in journals. A number of paintings depict snow and ice in places that never get that cold today—a nice bit of evidence, but hardly proof positive. After all, painters have been known to take artistic license. The canals in Holland did ice up almost every year; this hardly ever happens today. And what about the fact that during the 1600s, the Thames River routinely froze over solid enough for Londoners to stage "ice fairs" on it?

"From a modern perspective, the imagery [of ice fairs on the Thames] may appear romantic, but for the millions of people who relied on sunshine for good harvests it was terrible. With so many hovering at subsistence levels, food shortages during the Little Ice Age brought great hardship and suffering," writes science journalist Stuart Clark.[9] This is often cited as proof that the Little Ice Age really happened, because such deep freezes no longer occur. Then again, the Thames has been dredged since those olden days, making it deeper and swifter and a lot less susceptible to freezing over.

On balance, the mass of anecdotal evidence overwhelmingly indicates that the LIA was a time of chilly, damp summers and long, hard winters throughout much of Europe. The continent got so much rain that crops rotted in the fields, leading to famine,

plague, and ultimately the collapse of governmental power structures, which in turn resulted in bloody chaos—for example, the
Hundred Years War between England and France (1337–1453).
Of course there were many other factors contributing to societal disruption. Everything from the peccadilloes of the rulers
to Gutenberg's invention of the printing press in the mid-15th
century (and the subsequent mass production of Bibles and other
books) helped determine the course of history—events that had
nothing to do with the behavior of the Sun. Again, though, the
Moody Sun Hypothesis is not an exercise in environmental de-
terminism, but simply a way to factor the influence of the Sun's
changing behavior into our understanding of history. In the case
of the LIA, a significant and prolonged decline in temperatures
that accompanied the falloff in solar activity was especially destabilizing to the predominantly agrarian economy, the domino that
caused all the others to fall.

Unlike the relatively localized MWP, the LIA ran the entire
Northern Hemisphere, perhaps the whole world, through a
gamut of catastrophes. China and Easter Island give us a sam-
pling of the LIA's effects.

Although China, in the Northern Hemisphere, was hit with
drought, the opposite of what happened in rain-soaked Europe,
the end result was famine all the same. According to Sultan
Hameed, a solar physicist with the State University of New York
at Stony Brook, China's food production was devastated during
the first decades of the 17th century by a 15-year drought. Canni-
balism and insurrection arose in 1628, toppling the ancient Ming
dynasty in 1643.[10] Imagine if today's China, quickly becoming
the world's leading economic power, were once again faced with
a new solar minimum, and along with it another 15-year drought

and subsequent famine. Its 1.5 billion–strong society would drift into chaos, with geopolitical fallout rippling throughout the rest of the world. China's drop-off in agricultural productivity would cause shortages that today's world, with only about 30 days of food supply in hand, could ill afford.

The case of Easter Island vividly illustrates the effects of global cooling in the Southern Hemisphere during the LIA. It also dramatically recapitulates the current environmental debate. In *Collapse*, Jared Diamond singles out the remote Chilean territory, 2,350 miles west of San José, Chile, in the South Pacific Ocean, as a culture brought down by its own excesses. It serves as yet another cautionary tale for the wastrel world today. Diamond cites overpopulation and an overweening obsession with creating and transporting those gigantic Moai statues for which Easter Island is so famous, as the primary reasons that the isolated civilization laid itself low. What price art? The island was deforested and its economy collapsed under these burdens. Diamond clearly believes that the Easter Islanders made a mistake and should have chosen sustainability over immortality, a debate that is beyond the scope of this book.[11]

In either case, there may well be more to the story than the professor's tsk, tsk, tsk. The LIA's drop in temperatures and therefore in agricultural productivity may well have been the x-factor that tipped Easter Island's scales to empty. Grant McCall, an anthropologist at the University of New South Wales in Sydney, Australia, has spent 30 years studying the Rapanui people who inhabit the island, and he firmly believes that the era of global cooling was instrumental in halting that society's development. The island was cut off from contact with the even more frigid south for much of the year, hampered by fierce winter storms and frozen or impassable seas. Seafaring, and the fishing and the trade that accompany it, was severely curtailed. Moreover, McCall contends that the climate changes wrought by the LIA inflicted drought upon Easter Island's environment, impoverishing the populace. In short, he is disinclined to blame so much of the society's collapse on the natives' obsession with building their gigantic, mesmerizing statues.[12]

Beyond the particulars of that little civilization's plight, the Easter Island case study asks a broader question. To what extent are human beings responsible for what happens in their environment? Diamond's "nostra culpa" admonition against human profligacy seems more virtuous, yet also more egotistic, than McCall's fatalistic acceptance that some ecological disasters are simply beyond our control. Consider the fact that the Medieval Warm Period was followed by the Little Ice Age, which in turn was followed by our Current Warm Period (CWP), 1850 to the present. Are warming and cooling phases on Earth cyclical and therefore as unstoppable as the ocean waves? Is there an ineluctable rise and fall to climate history? The past thousand years have seen us warm up, cool down, and then warm up again. Does this mean that the next cooldown is on its way? Perhaps the current global warming is simply a restorative response to the LIA. Or has the burgeoning accumulation of greenhouse gases pushed what might have been only a simple, balancing correction into dangerous overcompensation? If that is the case, we will turn out to be Easter Island writ large, with our self-destructive worship, not of outsize Moai statues but of outsize consumption, hastening our fate.

COOLING VOLCANOES

The dirty little secret behind the Moody Sun Hypothesis is that levels of solar activity do not always match up as convincingly with conditions on Earth as happened during the LIA, or with the great global warming 12,000 years ago that melted the last great Ice Age. Case in point, the Medieval Warm Period, which saw temperatures rise without a corresponding increase in number of sunspots. The good news is that the Sun–Earth relationship is complicated enough to make a supercomputer weep; no system that has coevolved for almost five billion years can be expected to have simple, transparent mechanisms. For example, as noted in the preceding section, the lack of sunspots during the LIA caused some areas, such as Europe, to experience heavy rainfall, and others, such as China, to experience drought. Was this simply a shifting of weather patterns such that one region of the Northern Hemisphere received another area's precipitation along with its own? The mechanisms were undoubtedly finicky, involving solar winds, cosmic rays, cloud formation, ozone levels, and so on. In sum, one can always justify complex speculation in support of the sunspot–climate connection, thus making the case that cause-and-effect triggers and consequences do not happen in a timely fashion. Many a scientific argument has been won by, or at least decided in favor of, that person who best mastered the art of stultifying complexity; the old "baffle them with bullshit" approach, with a Ph.D. Recently, and refreshingly, a group of physicists from the University at Buffalo scored a major blow for clarity in understanding the sunspot–climate connection.

"For a long time people have tried to find out whether, for example, periods of maximum sunspots will influence the climate to behave in a certain way. . . . Whenever scientists thought they had discovered something, say they were seeing a positive correlation between temperature and sunspots, it would continue like that for several years and, all of a sudden, there would be a reversal and instead, they would start to see a negative correlation," said Michael Ram, University at Buffalo physicist and coauthor of the study.[13] It wasn't that there was no correlation between sunspots and global warming. Rather, the correspondences were strong until, suddenly, they reversed.

How can our atmosphere be so schizophrenic? Let's go back to the basics. Sunspots (with their heightened solar radiation) block out cosmic rays. Fewer cosmic rays mean fewer clouds, thus exposing the planet's surface to more sunshine and causing it to warm. Bottom line: an increase in sunspots yields an increase in Earth's temperatures—or that's the tidy little theory of how it all *should* work, which it does . . . in the upper atmosphere. Upper atmosphere temperatures do indeed fluctuate in pretty close correlation with sunspot activity, moving higher during solar maxima and lower during solar minima. That's what one would expect. However, changes in upper atmosphere temperatures are not always accompanied by corresponding changes in the lower atmosphere. The discrepancy is largely due to other temperature-impactful processes going on down here on the surface, such as greenhouse gas emissions, which raise temperatures, and volcanic eruptions, which lower them.

Volcanoes as air conditioners is a difficult metaphor to wrap one's mind around. That fire, steam, smoke, and cinders serve much the same function as a stream of chilled air is counterintuitive but nonetheless an apt comparison, because the ash and soot expelled into the air serve to block out sunlight and therefore depress temperatures. The cooling effect is particularly strong in the locales both where volcanoes erupt and where the prevailing winds carry their discharge. "By carefully studying the timing of other volcanic eruptions, we found that they coincided with all of the correlation reversals between sunspots and climate," said Ram.[14]

The research done by Ram and his coauthors suggests that high levels of solar activity might somehow trigger counterbalancing volcanic eruptions. Are volcanoes like great fountains of melanin, Earth's way of tanning itself to absorb and defend against spikes in the Sun's rays? No hard science of which I am aware comes close to proving or disproving such conjecture. However, it seems plausible that just as volcanoes powerfully affect the climate, climate might correspondingly affect the eruption of volcanoes. There appears to be a negative feedback mechanism whereby a warming global climate somehow triggers cooling volcanic eruptions. This would be hard to prove, however, because the presumed volcanic response could come centuries later. For example, the Medieval Warm Period coincided with one of the deepest lulls in volcanic activity recorded during the entire Holocene era. This lull would have had extra impact on Scandinavia, particularly Viking Iceland, normally a hotbed of volcanic activity. Perhaps this is why these countries are where MWP warming was most pronounced. With no volcanic ash or soot shading out the sunlight, temperatures rose apace.

Should not Earth then have erupted in volcanoes to cool things back down after the MWP? It did precisely that several hundred years later, spurring the LIA, which saw volcanic blowups reach record levels. A multinational research group led by the University of Colorado Boulder recently concluded that "an unusual, 50-year-long episode" of four extremely powerful volcanic explosions "triggered the Little Ice Age between 1275 and 1300 AD." Thus the Little Ice Age got caught in a double whammy, precipitated by a spike in volcanic activity and then extended by the Maunder minimum dip in solar activity.[15]

It is humbling to the contemporary human ego to acknowledge that just a handful of volcanoes around the world could offset global warming. How Lilliputian of us that just a few primitive, mindless eruptions could undo what took a century and a half of Industrial Revolution to cause! Not that we are proud, exactly, of all the pollution emitted by civilization, our masterpiece—just vaguely affronted that the resulting climate change could be counterbalanced by a few random mountain burps. Still, how wonderful if Gaia were to give us the gift of global cooling. Little Ice Age II, anyone?

5

When the Sun Fell Asleep

Edgar Cayce, the famous twentieth-century psychic healer and prophet, believed that sunspots were manifestations of spiritual and emotional conditions on Earth. A grand thought. Still others hold that our planet can cause the Sun to spot. Richard Michael Pasichnyk, a respected independent scientist, advances the theory that the Sun, because of its gaseous, almost gelatinous nature, is far more susceptible than planets to gravitational pulls, pulls that result in sunspots. But hardly anyone else seems to agree with these audacious thinkers.

We look up to the Sun. The Sun looks down on us, or would, if it could actually see anything. It has all the power and authority. We have none. What is lacking in this relationship is a sense of reciprocity, a feeling that what we earthlings say and do registers with the Sun in any way: Sol as oblivious monarch, unfeeling god. There is, however, one shining exception to the Sun's otherwise pristine record of aloofness, one time when it might actually have responded to us humans, when *we* controlled *its* behavior and not the reverse.

It tempts the believer in us all to note that the reign of the Sun King, France's Louis XIV, 1643–1715, almost exactly coincides with the Maunder Minimum, 1645–1715, a 70-year period during which sunspots vanished almost completely. Only a tiny handful were observed over those seven decades, an inexplicable absence confirmed by the almost complete lack of residual radioisotope evidence, which would remain today had more sunspots existed but just gone unnoticed. There is no known historical analogue for this amazing dormancy; it is represented by one of the lowest points of Usoskin's solar graph. The fact that sunspots, which normally ebb and flow in size and frequency over 11-year cycles, just up and disappeared for more than six whole solar cycles was not investigated in any scholarly journals until the late 1800s, when Edward W. Maunder, the English astronomer after whom the Maunder Minimum is named, retroactively pieced it all together. His report then proceeded to gather dust for almost a century, until 1977, when Jack Eddy, an astronomer with the High Altitude Observatory in Colorado, picked it up.[1] Eddy grasped that Maunder had documented an astonishing and unprecedented disruption of the solar cycle. Eddy was also the first one to note the near-perfect overlap between the Sun King's reign and the Maunder Minimum.

Royalists might have deemed it only fitting that the Sun should clean up its act during the reign of *Le Roi Soleil*, the Sun King, if not out of deference, exactly, then at least as a professional courtesy, one heavenly luminary to another. The Sun's obedient display of spotless perfection while Louis was on the throne was just added proof that God had ordained Louis with kingly greatness that extended beyond Earth and into the celestial realm. The basic complaint against these unseemly blotches was that they violated the Church's doctrine of divine errorlessness, the aforementioned sacrosanct Aristotelian notion that because God had made the heavens, they were therefore perfect. Unsightly sunspots were either ignored or explained away as the transits of planets, or as space junk casting shadows on the face of the Sun, or as motes in astronomers' eyes or imperfections in their telescope lenses. Or worse, some heretical nonsense cooked up by Galileo. Something needed to be done, and that something was the construction of the Sun King persona.

According to Ellen McClure, in her scholarly history of the reign of Louis XIV, "Galileo's sunspots brought the celestial sun into the realm of terrestrial corruption. Against [this] development, the restoration of faith in an order grounded in permanence and transcendence was more urgent than ever before. Efforts to effect this restoration resulted in the creation of the Sun King, whose authority and monarchical identity was constructed at least in part to counter the destructive implications of the sunspots." At first, McClure's argument seems like academic artifice, one of those after-the-fact reconstructions that professional intellectuals are paid to get off on. Did French folks really care that much about sunspots back then? Enough to cast their ruler as some celestial Mr. Clean? What to us seems much ado about nothing, McClure pegs as a high-stakes political gambit: "Considering sunspots and the Sun King together not only reminds us that they belong to the same period. It also reminds us of the cosmic stakes involved in restoring and recasting political authority."[2] Whoever succeeds in claiming the mantle of the Sun must therefore reign supreme.

Right from the start, there was something supernatural about the boy who would become king. Louis XIV was born in 1638, 23 years after his estranged and childless parents had set out to create an heir to the crown. His quasi-miraculous appearance (in the same year that Galileo went blind) earned him the nickname *Dieudonné* ("God-given") and was taken by many as evidence of divine intervention. In 1643, at the age of five, he assumed the throne under the regency of his mother, Anne of Austria. Whether it was she, Louis, or one of their advisors who came up the Sun King conceit, no one knows for sure. But by all reports, the boy took to his role with a special gusto. He particularly liked to dress up as Apollo, the ancient Greek Sun god.

Louis XIV quickly learned to cast himself as the Sun's human embodiment, a useful comparison, for what could be superior to the Sun except God? Louis branded himself as the Sun King as consciously and relentlessly as the Dallas Cowboys have branded themselves as "America's Team." Drawing pithy parallels between Sun and monarch became something of a parlor sport in Louis's court. Both move inexorably and stray not from their preordained course. Both provide manifold blessings, most of which go unnoticed by the (ungrateful, peasant, low-life) recipients. Extra brownie points for flattery, such as pointing out that the Sun is mindless, while the king acts with will, intelligence, and grace.

"Even this delicate matter of the Janus-like king, whose gaze is at once fixed upon God and upon his subjects, lends itself to characterization by the sun, whose glory reflects that of its creator, all the while granting its beneficial rays to the people below," observes McClure.[3]

Any aspersions cast on the Sun reflected directly on the king himself, and were therefore to be denied. Theories that sunspots were not optical illusions or objects extraneous to the Sun, but rather part of the Sun itself, threatened to undermine and invalidate divine right political theories connecting the king to the Sun and to God himself. Seventeenth-century French court philosopher Pierre LeMoyne dismissed sunspots as imperfections in the people's ability to fully behold the grandeur of the king. "Let astrologers reproach it for having spots that it does not have; let them accuse it of sterilities of which it is innocent: let poets make up stories of its loves and its gallantries: let them give it Mistresses and bastards: let others charge it with the birth of serpents and poisons, the sun will not distance itself for all that: it will not fail to shine on them," wrote Le Moyne.[4]

Were people another species back then? To us it seems such a wild stretch of the imagination that a few spots on the Sun could somehow shake the fundaments of society. Part of what was so upsetting about Galileo's *Letters on Sunspots* was his insistence that sunspots were neither illusions nor extraneous, but rather were integral to the Sun's essence. Kind of like the solar version of original sin. To the Renaissance mind, the idea of the Sun being composed of discrete and mismatched particles "undermined the concept of cohesion necessary to support the doctrine of transubstantiation,"[5] a Roman Catholic orthodoxy, which holds that the bread and wine received during Holy Communion are at that moment quite literally the body and blood of Jesus Christ. The Church further regarded the sunspot heresy as a serious political blunder because it played into the hands of the Protestants, who downplayed Communion as merely symbolic, meaning that Christ's body and blood were consumed only metaphorically. Such sunspot quibbles might to us seem get-a-life farfetched until we consider that solar blemishes do, in fact, threaten contemporary civilization—not metaphorically, but flesh-and-blood physically—with the plasma storms that periodically issue from them. Perhaps our nonchalance toward sunspots is ultimately less sane a response than the hubbub that gripped our ancestors.

A pity that Louis XIV never learned firsthand that the sunspots actually vanished during his reign. Although war, intrigue, and amorous dalliance far outranked astronomy among the Sun King's interests, His Majesty would nonetheless have been pleased by the Sun's spotless homage. But the fact of the matter is that as sunspot activity ebbed during the Maunder Minimum, so did scholarly interest. The Church didn't like the blemishes one bit and was quite content to see them fade. Scientists saw the spots as distant solar curiosities and little else, their absence providing little opportunity to study how they might affect our life here on Earth. Doubtful that either group saw any connection between the absence of sunspots and the fact that the Sun King had one of the coldest, dreariest reigns in history. At the Chateau Versailles, carved golden images of the Sun festooning the wrought-iron gates were meant to reflect the sunlight so brightly that one was forced to look away, as appropriate for mere mortals beholding the face of the deity. But no one did much squinting during Louis's reign, when cloudy days so outnumbered sunny ones that the palace's brilliant gold trim and burnish might just as well have been dull bronze.

With the passing of Louis XIV, the sunspots returned and Sol fell in our estimation, the way aging fathers so often do. As for the question of whether Louis XIV had anything to do with the disappearance of the sunspots during his reign, 21st-century scientism insists, disappointingly, that it was all just a coincidence after all. Too bad, because we could really use a superhero Sun King right about now.

PRESENT

My earliest memory of the Sun is of it tasting good. I was about three and a half, summer 1957, in the garden behind our apartment on a dirt road in Danbury, Connecticut. The landlord, a kindly man, was fussing over his tomato plants. He told me to pick a ripe red one off the vine and take a bite. Warm and delicious! I could taste the sunshine right inside it.

I like to think that the summer of 1957 brought a bumper crop of Sun-related pleasures to people everywhere. Not that 1957 was meteorologically remarkable; it was not. It was, however, truly exceptional in terms of our relationship with the Sun. Human evolution changed fundamentally and forever with the launch of the Sputnik I satellite on October 4, 1957. Even though the Russian satellite mission had nothing to do with the Sun per se, the fact is that it inaugurated the Space Age, which quickly caused the sum of our scientific knowledge about the Sun to skyrocket, quite literally.

Over the past half century, our understanding of the Sun has undergone a wholesale transformation. True, Space Age solar research has served to recontextualize our star as just one of a teeming cosmic crowd, but what the Sun has lost in status it has gained in cache. What was long assumed to be just a massive, plodding power source is now reconceived as a complex and dynamic broadcaster of energy and information, a potent source of both health and disease. In this section, we will see the Sun in a new light, as our silent partner, sometimes supportive and other times treacherous, and never to be underestimated for the influence it has on our lives.

6

Sunspots Are Making the World Warmer

"For some years now, an unorthodox idea has been gaining favor among astronomers. It contradicts old teachings and unsettles thoughtful observers, *especially climatologists* [italics mine]. The sun is a variable star," explains Lika Guhathakurta, chief program scientist of Living with a Star, the international solar physics awareness program headquartered at NASA in Washington, D.C.[1]

"Bullshit!" says Al Gore.

I admire Al Gore, vote for him every chance I get. Once, in Washington, D.C., I heard him give a perceptive, heartfelt, and hilarious talk about the Middle East. Who gets laughs out of Arabs versus Jews? No, he did not invent the Internet, although a science buddy of mine who attended a top-level information technology roundtable with Gore said he had the sharpest mind in the room.

With regard to climate change, I believe that Gore is on the side of the angels, as well as the facts, mostly. And for the record, his above-noted epithet was not hurled at Dr. Guhathakurta directly. Rather, the tireless public servant took on the general category of scientists who believe that changes in the Sun's behavior are connected to changes in Earth's climate.

As Gore, in a tirade at the Aspen Institute in August 2011, explained things:

> There's no longer a shared reality. . . . [I]t's no longer acceptable in mixed company—meaning bipartisan company—to use the goddamn word "climate." . . . And some of the exact same people—I can go down a list of their names—are involved in this. And so what do they do? They pay pseudo-scientists to pretend to be scientists to put out the message: "This climate thing, it's nonsense. Man-made CO_2 doesn't trap heat. It may be volcanoes." Bullshit! "It may be sunspots." Bullshit! "It's not getting warmer." Bullshit![2]

As one who concurs wholeheartedly with Guhathakurta's contention that the Sun's variability affects our climate, it pains me considerably to be bullshit-lumped with two other opinion categories—"man-made CO_2 doesn't trap heat" and "it's not getting warmer"—to which I most definitely do not subscribe. Neither do I cotton to being accused of having paid pseudo-scientists to concoct some obstructionist malarkey—Lord knows I have better things to do with my money—or of being one of the concocters myself. For the record, none of my work on this book or any of my four preceding books, *Gaia: The Growth of an Idea* (1986),

Common Sense: Why It's No Longer Common (1990), *Apocalypse 2012: An Investigation into Civilization's End* (2007), and *Aftermath: Preparing for and Surviving Apocalypse 2012* (2010), was funded by any individual or group other than the publishers of record, plus sundry cash advances from my mother, my (ex-)wife, and the Visa Corporation of America, usually at 19.8 percent. If I am a tool of the climate change naysayers, I am quite a dull one, because I have never received a dime from anyone seeking to influence what I write regarding the Moody Sun Hypothesis or any other topic. I therefore beg respectfully to differ with the esteemed former vice president.

How did we arrive at the point where it is deemed ludicrous, even seditious, to suggest that the Sun, which provides all our light and warmth, might also be partly responsible for the global warming we are now experiencing? The tactic of denouncing as illegitimate and dishonest anyone who questions the central tenet that climate change is overwhelmingly anthropogenic is destined, like all dictatorial fiats, for the ash heap of history. For the record, I am not a climate change denier. I believe that the planet is indeed warming significantly, quite possibly dangerously, and that that warming is largely, though certainly not exclusively, attributable to the emission of man-made greenhouse gases such as carbon dioxide and methane. What I denounce, deny, and reject is the totalitarian impulse to mute discussion on this subject, to proclaim the scientific debate to be over, as Al Gore's climate reality project attempted to do with its 24-hour onslaught of global programming in September 2011. We are just now on the verge of attaining a true working understanding of the role that the Sun plays in climate change. As noted earlier in this book, the International Heliophysical Year (IHY) 2007–2008, which inspired some of the greatest collaborations in the history of science, saw unprecedented numbers of profoundly sophisticated spacecraft sent aloft to examine the Sun and the Sun–Earth relationship. The data needed to evaluate the Moody Sun Hypothesis are now streaming in. Analysis and interpretation will continue for at least the next decade. To close the book on the Sun now would be the height of arrogance and the depth of idiocy.

Sunspots do very much belong in the climate change debate. They are the best historical record of solar activity we have and are relevant to the warming and cooling cycles our planet has experienced throughout the ages. Flip back to Usoskin's graph. See those two giant spikes all the way to the left, between 12,000 and 10,000 years ago? As explored in a preceding chapter, those dramatic outbursts of solar activity accompanied the powerful global warming that reversed the most recent Ice Age. Common sense says that this Sun–Earth parallel is more than just a coincidence. Ditto the giant spike at the other end of the graph, a sunspot surge occurring over the past century and a half or so. Sometimes the Sun just gets in the mood to spit radiation at us. That's all there is to it. The recent barrage of solar activity has without doubt contributed significantly to the rise in temperatures, though probably not so powerfully as the accumulation of man-made greenhouse gases in the atmosphere. *Neither* factor can be ignored.

The most lethal greenhouse gas, judged in terms of negative impact on the commonweal, is the hot air that spews from the mouths of ideologues in the climate change debate. Before launching his "bullshit" tirade, Gore would have done well to consider the wise words of George H. W. Bush: "I remind myself

a lot of this: We must conquer the temptation to assign bad mo-
tives to people who disagree with us."[3]

"Bullshit!" does not quite capture that spirit of open-mindedness.
Gore's Calvinist zeal to triumph over the devil that is greenhouse
gas precludes him from considering any other explanation. As
Earth's climate changed and temperatures rose over the past
century and a half, our only significant source of heat and light
has been largely ignored as a possible cause. The Sun and global
warming? No relation . . .

The problem here is not just a lack of civility, but a lack of
understanding that the greenhouse–sunspot debate over global
warming is not just a *scientific* one. It cuts way deeper than that.
Despite the mountains of data from satellites, research expedi-
tions, computer models, and laboratory experiments, at bottom
the climate change debate has devolved into an ideological feud
about the role human beings play in the world. The bluenoses
from Gore's "nostra culpa" side of the aisle stress how important
and therefore culpable human beings are in the grand ecological
scheme of things. They remind us that with great power comes
great responsibility, so the burden of correcting any environmental crisis, such as climate change, falls to us. The naysayers, those
who minimize the importance of climate change and/or the
anthropogenic component of it, see things differently: we human
beings are not the drivers; we are just along for the ride.

PAVED WITH GOOD INTENTIONS

Gore is the front man for establishment climate science. His
colleagues at the Intergovernmental Panel on Climate Change

(IPCC) are impressive for their intellectual and professional achievements as well as for their dedication to the global commonweal. Founded in 1988 by two UN organizations, the United Nations Environment Programme and the World Meteorological Association, the IPCC focuses the efforts of more than 2,000 climatologists conducting research in at least 150 countries, most volunteering their time and effort to the common goal of issuing definitive reports on global climate change. IPCC members are organized into three working groups: physical science, environmental and human impacts, and policy options. The clear consensus of the IPCC is that climate change is indeed taking place, largely in the form of global warming. And the most important cause, by far, is human activity, particularly the emission of greenhouse gases such as carbon dioxide and methane. Climate change, already causing dangerous environmental impacts ranging from unprecedentedly severe storms to burgeoning desertification, must be slowed or, better yet, reversed by curbing the accumulation of these gases, lest millions, perhaps billions of people be imperiled as temperatures rise and quality of life continues to degenerate—that's the gist of the IPCC's philosophy.

Before Gore's "bullshit" outburst, the rhetoric in the climate change debate had reached the "After you, Alphonse," "No, after you, Gaston," phase of feigned civility. Naysayers rarely failed to acknowledge the importance of greenhouse gases, and bluenoses usually managed to doff their hat to venerable old Sol. Judith Lean, a leading solar physicist with the United States Naval Research Laboratory and an ardent IPCC exponent, still observes the niceties. "'Solar constant' is an oxymoron. Satellite data show that the Sun's total irradiance rises and falls with the sunspot cycle by a significant amount. Understanding solar variability is

crucial. Our modern way of life depends upon it," she offers.[4]
Judging from this sound bite, Lean could be counted as a supporter of the Moody Sun Hypothesis—except, that is, when it
comes to the subject of climate change.

Women are few among solar physicists, so they stand out—especially Lean, who, when I met her in 2005 at a conference sponsored by the University of Colorado's Laboratory for Atmospheric
and Space Physics, had the looks of Hepburn, Audrey, and the
commanding air of Hepburn, Katharine. Lean cochaired the
conference that, as described in my previous book, *Apocalypse
2012*, just happened to coincide with one of the wildest weeks
ever recorded on the face of the Sun, mammoth sunspots busting
out all over its face. Even though the massive solar outburst came
immediately after Hurricane Katrina and before Hurricanes Rita
and Wilma (both storms larger than Katrina), the eruption was
never once mentioned among the 100 or so solar physicists there
gathered, neither in their presentations nor even, to the best of
my knowledge, during the coffee breaks. Megastorms on the Sun
coinciding with megastorms on Earth—might there possibly be
a connection? Lean coolly ignored the distraction and gave a
fact-packed literature review of the role of "solar forcing," jargon
for the influence of solar variation on Earth's climate, concluding
that the Sun's total output rises and falls so minutely—only about
0.1 percent from solar maximum to solar minimum—that its role
is of minimal importance to climate change. Perhaps the IPCC's
position on solar forcing ultimately depends too heavily on Lean,
who holds the position of IPCC's lead author on the issue; as of
2010, Lean had selected for publication only those papers which
she had authored or coauthored. This is not science. This is
ideology.

Ironically, Lean's dismissal of the Sun's role in climate change flies in the face of the fact book provided at the conference she cochaired: "Some empirical models estimate that the sun has varied by nearly 0.5% since pre-industrial times. Climate models indicate that such a change may account for over 30% of the warming that has occurred since 1850."[5] Over 30 percent! That's not negligible.

I would love to debate Al Gore on this subject of the Moody Sun Hypothesis, but real life isn't *Rocky* and the champ doesn't stick his chin out to everyone who wants to take a punch. So in tribute to Gore's oratorical excellence—remember when shortly after being elected vice president he went head-to-head with H. Ross Perot and pretty much knocked that guy out for good?—I propose a debate between Lean and Harvard-Smithsonian astrophysicist Willie Soon, who, like Lean, politely remembers to give the mandatory nod toward his opponents' position:

"It is fair to say that the scientific inquiry thus far has established the enormous complexity of the 20th century warming problem. The two major drivers appear to be the varying brightness of the Sun and, of course, the accumulating greenhouse gases," writes Soon, an ardent naysayer who believes that solar variability is more in the 0.5 percent range.[6] Soon has long argued that a significant share of current global warming is due to the increase in sunspot activity from the mid-1800s through around 1980. Thereafter, greenhouse gas accumulations took over as the principal driver of climate change. Debating tip to Lean: Soon will argue that a stronger Sun evaporates more water from the oceans. He will remind you that water vapor, which outweighs carbon dioxide by many orders of magnitude, accounts for the greatest share of warming in the atmosphere. Debating tip to Soon: Lean knows that you receive an average of $100,000 per year from petroleum interests.

SELFISH BILLIONAIRES

Gore claims that selfish billionaires are paying off scientists to pooh-pooh the whole idea of man-made global warming. This must be true, given the oil and gas cartel's trillion-dollar stranglehold on the status quo. Indeed, Soon's major funders include ExxonMobil and the Charles Koch Foundation, which is supported by the right-wing oil, gas, and chemical tycoon for whom it is named. So on the bluenose side we have collusion and on the naysayer side we have greed. Both vices make it tempting to fudge results, as happened in the case of the East Anglia Climatic Research Unit (see "Climategate" section below). It is my considered opinion, however, that the majority of scientists on both sides of the debate are largely motivated by their passion for knowledge, that they truly hold the beliefs they espouse, and that those beliefs guide them in pursuing appropriate sources of assistance. Just because a billionaire, or a former vice president, supports your work does not mean that you have to deliver the results that supporter wants. Of course, a spoonful of sugar does help the facts go down a little easier.

If I were cadging for a bribe—uh, research grant—from naysayer fat cats to help fund my Moody Sun Hypothesis research, I would hammer home the point that the planet has alternately warmed and cooled throughout its history, with and without human input, and that "what goes up must come down" applies also to global temperatures. Much as your body adjusts itself, sweating to cool down when it gets too hot, and then shivering to warm up when it gets too cold, so does the climate regulate itself, in a process known as homeostasis. Of course, the time-scales are vastly different for bodies and planets—seconds versus centuries—as are the mechanisms; volcanoes, as noted earlier, are one of the ways that Earth has of cooling off. Whatever the mechanism, the rise and fall of global temperatures over time is as undeniable a reality as the ebb and flow of the tides.

For good measure, I would throw in the appealing yet scientifically tenuous argument that climate change can be good for you, and that much of our fear of global warming is actually just fear of the unknown. It's sure as hell better than our starting on another Ice Age, with vicious winters, cold and moldy summers, famine, and plague. What's so bad about milder winters anyway, especially in the more northerly latitudes of North America, Europe, Russia, and China? In *Aftermath*, I wrote of enduring a weeklong heat wave in Siberia, where global warming is nonetheless more popular than chocolate. To be fair, though, rising temperatures are not so good for the teeming Third World millions who live in equatorial desert regions.

Besides, I would argue, there's nothing much we can do about global warming. Who *really* believes that the world, with its burgeoning populations and economies, will actually cut back sufficiently on oil, gas, and coal consumption (and the resulting greenhouse gas emissions) anytime soon? "Relax, humankind will find a way to adjust. We always have and we always will," or some such, would be the refrain of my pitch to be put on the nay-sayer payroll. On the bright side, global warming is sure to trigger a cool, relaxing Ice Age (though probably tens of thousands of years from now).

If, on the other hand, any of that million-dollar Nobel Prize money is still burning a hole in Gore's/IPCC's pocket, I would press the argument that what's most confusing about the Sun's alleged role in global warming is that the wavelengths don't match up. The proton radiation that typically issues from solar blasts has a much shorter wavelength than the infrared radiation that yields heat. Warming up the climate with proton injections is about as efficient as cooking a stew over a lightbulb—that bulb would have to be awfully bright just to cause even a few bubbles. Of course, if the lightbulb were left on for a month, say, whatever was left in the pot would probably be cooked through and through. The same idea holds for solar output and global warming: solar blasts may not do much individually to warm the climate, but cumulatively they do take their toll.

I also offer my services as author of *Common Sense: Why It's No Longer Common* (1994) to administer commonsense shock treatments to the Democratic Party, of which I am rapidly becoming a former member. À la Gore, the Democrats have chosen in a number of their climate change communiqués to belittle the possibility that sunspots and solar variability could in any way be relevant to global warming. How foolish and shortsighted, like slackers at the back of the classroom making animal noises. Paul Begala, a veteran Democrat apologist, took the following swipe at Ron Johnson, Republican candidate for the U.S. Senate from Wisconsin: "Ron Johnson doesn't think burning fossil fuels—or any human activity, for that matter—causes global warming. He blames sunspots. Yep, sunspots. I think Ol' Ron has been out in the sun too long."[7]

I think Begala has been under his desk too long. Johnson's not believing that any human activity contributes to global warming does indeed place him on the naysayer fringe. But his position that increased solar activity might have something to do with climate change is eminently reasonable. Ridiculing that notion, as Gore did with his "Bullshit!" tirade, backfires embarrassingly, because the Sun–Earth connection makes sense to people. Turn up the heat lamp and things get hotter. The argument that increased activity of the Sun, the ultimate source of virtually all the heat and light this planet has ever known, could somehow be related to global warming is not only deeply scientific, it is also quite plausible on its face. Don't mock common sense! PS: In November 2010, "Ol' Ron" Johnson unseated Senator Russ Feingold, the venerable three-term Democratic incumbent.

To be fair, the Gore/IPCC contingent bears an unfair burden of proof. Not only must they demonstrate that human activity is causing climate change, and that such change is harmful to us, but they have to do so convincingly enough to inspire us to take painful action ASAP. In order to win popular support for doing battle with the behemoth energy monopolies, for imposing new regulations and restrictions on the faltering economy, and for living lives that, at least temporarily, are of diminished comfort and convenience, climate change exhorters must make grave danger seem imminent. The last time Americans felt roused to action against danger was in uniting against the terrorists responsible for 9/11. Not that we all are looking for a way out, exactly, but if enough doubt is cast on the whole climate change thing— well, it's kind of tempting to postpone taking action until the facts are crystal clear. There are, after all, other pressing problems in the world today—the hateful religious conflict among Muslims, Christians, and Jews, illegal drug trafficking and the savage violence that results, the looming threat of global economic depression—that don't require us to turn down the air-conditioning.

RIPPING A NEW O-HOLE

The bluenose campaign to curb greenhouse emissions is bolstered by a solid success story: the control of another set of man-made gases—in this case, CFCs (chlorofluorocarbons), which deplete the stratospheric ozone layer that protects us from excessive exposure to potentially carcinogenic ultraviolet rays. English atmospheric scientist James Lovelock discovered in 1972 that CFCs were accumulating worldwide, released from such ordinary sources as refrigerants and aerosol spray propellants. In 1974, F. Sherwood Rowland, Paul J. Crutzen, and Mario J. Molina worked out how these aerosols—gaseous suspensions of fine solid or liquid particles—do indeed rip holes in the ozone layer, work that led to their sharing the 1995 Nobel Prize for chemistry. In 1989, the international Montreal Protocol on Substances That Deplete the Ozone Layer went into effect. The treaty had the goal of reducing and eventually eliminating the emission of CFCs. In most cases, these gases were replaced with far less corrosive HCFC (hydrochlorofluorocarbon) aerosols. The international agreement has worked: CFC emissions have plummeted and the ozone layer has begun to mend.

Today, the preponderant majority of atmospheric scientists express the same level of confidence that man-made greenhouse gases are warming the atmosphere. Much as removing man-made CFCs helped correct a grievous problem in the atmosphere, so must we take concerted action to reduce the man-made greenhouse gases playing hob with our climate; that's how Gore's/IPCC's unexceptionable reasoning goes. The difference is that removing CFCs caused little economic hardship. Other than the mandated reformulation of some spray aerosol products and a new set of regulations about the disposal of Freon™ obtained from decommissioned air conditioners, refrigerators, and freezers, the transition from CFCs to more ozone-friendly aerosols was barely noticeable. If stemming carbon dioxide, methane, and other greenhouse gas emissions by switching from fossil fuels to renewable ones were so painless, the changeover would probably have happened a long time ago.

Naysayers can prick holes in the ozone/greenhouse gas analogy. For example, the largest Antarctic ozone hole ever recorded came in autumn 2006—well after the CFC ban—spanning an area slightly larger than North America. The largest Arctic ozone hole came even more recently, in late winter 2011. Though significantly smaller than its infamous Southern Hemisphere counterpart, the 2011 hole was far and away the worst one ever recorded in the Northern Hemisphere. Atmospheric scientists insist that the major reason for these setbacks in the recovery of the stratosphere is that CFCs take an average of more than 40 years to decay. Since the total amount of these chlorine-carrying aerosols peaked in 1995 (the peak came so late partly because of compliance problems with the Montreal Protocol), CFCs will likely continue to shred stratospheric ozone molecules well into the middle of the 21st century. Bottom line, no one seriously disputes that without the CFC ban, stratospheric ozone depletion would have been a lot worse than it is currently. Still, there's no avoiding the feeling that the emergence of record-setting ozone holes as recently as 2006 and 2011 does run contrary to expectations engendered by the CFC-emissions ban. Could it be that the Sun got into one of its moods? Sunspot activity was peaking at or near historic levels through the record-setting ozone depletion of 2006. The Sun then went nearly dormant for the next three years, rebounding just in time for the record-breaking ozone hole of 2011. "The total global amount of ozone becomes enhanced, depleted and enhanced again by 1 to 2% as solar activity goes from its maximum to its minimum and back to its maximum every 11 years. This modulates the protective ozone layer at a level comparable to human-induced ozone depletion by chemicals wafting up from the ground," writes astronomy professor Kenneth R. Lang.[8]

Like CFCs, sunspots can rip us a new o-hole by pummeling the atmosphere with CMEs (coronal mass ejections). CMEs are multi-billion-ton explosions of high-energy protons that billow out from the Sun and quickly rival its size, causing shockwaves that in turn create convoluted magnetic fields and that propel countless subatomic particles at near light speed. Some CMEs, particularly those that launch from the northwest quadrant of the Sun's face, hit Earth, where normally they are channeled by our planet's magnetic field to enter the atmosphere at the poles. The invading hyperprotons proceed to ionize oxygen and nitrogen molecules, which then fuse together to form nitric oxide. The nitric oxide reacts with ozone, O_3, a molecule composed of three oxygen atoms, and rips it into two parts, a conventional oxygen molecule of two oxygen atoms, O_2, and a third, loner oxygen atom. Such depletion is injurious to us humans, because ozone normally absorbs solar radiation at wavelengths between 200 and 300 nanometers, the heart of the ultraviolet (UV) spectrum. Every time an ozone molecule is destroyed, the likelihood increases that harmful UV radiation will penetrate to the planet's surface and burn someone's skin, sometimes triggering skin cancer. (See chapter 8: "Twenty-First-Century Sun Worship.")

CME-induced ozone holes usually close up in a matter of weeks, having reduced our UV protection levels by a percentage point or two, at most. After that time, the nitric oxide, a compound that when dissolved in water vapor becomes nitric acid, is washed out of the atmosphere by rainfall as acid rain. Ozone molecules then reform and resume their normal business of absorbing UV rays. No biggie. But every now and then the situation turns deadly. Studies of chemical deposits in ice cores mined from Greenland indicate that the CME of 1859, probably the most powerful solar blast to hit the planet in the past century and a half (more on this blast, known as the Carrington event, in chapter 11), blew a 5 percent hole in our planet's stratospheric ozone layer, comparable in size to the one opened up over the poles more recently by CFCs. The 1859 ozone hole is believed to have remained open until 1863. Consider that CFCs had not yet been invented back when the mammoth Carrington blast hit the atmosphere. By contrast, countless tons of these durable aerosols remain in the atmosphere even though almost a quarter century has passed since the Montreal Protocol banned their emission. Could not a deadly synergy arise between these lingering ozone slayers and a killer CME? Not according to bluenoses who claim that the relationship between climate change and sunspots is all BS.

CATASTROPHES AND STATISTICS

The climate change naysayers have an even bigger hole in their argument. Their claim that climate change is not actually happening sounds more and more preposterous with each passing catastrophe. Too many megahurricanes, freak tornadoes, record blizzards, and devastating floods to ignore. Too many trailer parks flattened, infernos triggered, main streets drowned. Gore/IPCC

definitely has the debating advantage here, with a corner on an-
ecdotes that, though sometimes iffy as hard scientific evidence,
have the ring of truth. Why, the melting stories alone—Antarctic
icebergs calving like rabbits, the Arctic ice pack going bald—are
enough to make one's blood run cold. Heat rises, so it stands to
reason that a warming planet would see its mountain ice caps
melt. And so indeed they have, from the fabled Kilimanjaro in
Tanzania to the South American Andes, and now, to Everest,
crown of the Himalayas. In 2011, Apa Sherpa retired from mountain climbing after scaling the world's pinnacle a record 21 times.
He explains that Mount Everest has become more dangerous
because ice that he had always counted on to be firm beneath his
feet is now melting, slippery, and unstable.[9]

The weather certainly does feel like it has changed. But can
we trust those feelings? Remember how big those snowdrifts
were when we were kids? Of course, we were only four feet tall
back then. Summers were always stinkin' hot, especially before
we finally got an air conditioner. Sure, disasters were always bad,
but nowhere near as frequent or horrendous as they are today—at
least that's the way it seems, judging from what we see on the
news channels flooding the television with HD color images
from every corner of the world every hour of every day. Me, I
can't decide if today's climate is out of tune or just transposed for
a new instrument, like a Bach piece played on the piano rather
than the original harpsichord.

Uncertain of our instincts, we are forced to march to the
drumbeat of climate change statistics. During the summer of
2011, both Oklahoma and Texas set national high-temperature
records. The month of July 2011 in Oklahoma had the highest
average temperature, including day and night, for any month, of

any state, ever, at 89.1 degrees Fahrenheit, up a full degree from the previous record, also set in Oklahoma, during the drought of 1954. Texas set the June–August day and night heat record at 86.8 degrees, along with a deluge of records for fire and flood during that same year. (Ironically, both are oil and gas states, prime naysayer territory.)

In all, well over 10,000 high-temperature records were reported nationwide to have been tied or broken during 2011. But no matter how huge the numbers pile up, there will always be that lingering, Mark Twain suspicion that there are "lies, damn lies, and statistics." Skeptics who hear that all these records are being set in Oklahoma, Texas, and wherever wonder how we can really know for sure. Were folks standing around looking at their thermometers during, say, the Dust Bowl of the 1930s? Could it be that we are setting so many records just because there is so much data around today? Megapixels and gigabytes pour out of every electronic orifice, and sometimes it seems that the only way to get anyone to give a damn about any of it is by claiming that some new mark has been set—for example, that Texas June–August heat record. While it may be true, as the saying goes, that the unexamined life is not worth living, that doesn't mean that every single aspect of it must be digitized. Comparing today's shiny new number sets to whatever ciphers were jiggered up back during the Depression is like matching up apples and crab apples; they are indeed comparable, but it's the fresh, tasty ones that usually make it into the pie. And all these quintillions of data points add up to less than 1 degree Celsius increase in temperatures around the world over the past 100 years? How would we have kept track of all this without supercomputers? Would we even have noticed the change? I hope no one dropped a decimal point!

In 2009 renowned Berkeley astrophysicist Richard A. Muller—remember, he's the one who highlighted the importance of McIntyre and McKitrick's debunking of the climate change "hockey stick"—decided to try his hand at settling once and for all the question of whether temperatures are really rising rapidly today. He obtained $600,000 in funding from a variety of sources, including $150,000 from the Charles Koch Foundation (which, as noted, typically supports climate change deniers). Muller's team examined two main criticisms made by climate change skeptics: that historical weather and temperature readings are unreliable, and that urban metroplex "heat islands" upwardly skew the data since a disproportionate share of measurements are taken in these locations.

After two years of research, in late 2011 Muller and his team declared that mainstream IPCC climate change scientists are correct: land temperatures have risen 1 degree Celsius (1.6 degrees Fahrenheit) since the 1950s. "The skeptics raised valid points and everybody should have been a skeptic two years ago. And now we have confidence that the temperature rise that had been previously reported had been done without bias," said Muller.[10] Reactions were predictable, ranging from "I told you so" from the Gore/IPCC camp to veiled expressions of betrayal by the skeptics; the Charles Koch Foundation sniffed that Muller had examined only temperature fluctuations on land, not on the ocean, and that research was ongoing. True to his role as honest broker, Muller left all sides of the debate a little bit unhappy. While he confirmed the incremental uptick in land temperatures, he still is not sure that greenhouse gases are the sole or primary cause of that warming, and he makes a point of not committing to that particular orthodoxy.

CLASH OF THE MAGNETIC FIELDS

Grasping the totality of the solar system's energetic interplay is not for the faint of heart or mind. Richard Michael Pasichnyk, an indefatigable independent researcher of voluminous output whom we met in the previous chapter, directs our attention to the dynamics of the solar system in order to help explain climate change here on Earth. He has spaded up dozens of correlations, coincidences, and connections and firmly believes that, all told, his research amply demonstrates that the Sun's behavior actively influences, even controls, our climate. This perspective jibes well with the Moody Sun Hypothesis. Pasichnyk notes the following:

Solar cycles are correlated with sea level, atmospheric pressure, and surface air temperatures in summer, and especially over the oceans in winter. Ozone varies with the long-term solar cycle, as do upper atmospheric airborne particles (stratospheric aerosols) and shifts in climate. The extent of Newfoundland's ice cover for the period between 1860 and 1988 has been correlated with solar activity. . . . The 11-year solar cycle is correlated with air temperature, air pressure, droughts, floods, lake levels, snowfall, tree abundance, and tree ring growth. Rivers such as the Nile, Ohio River, and Parana River (Buenos Aires, Argentina), rise and fall in accord with solar activity. Numerous examples exist that reveal an 11-year solar cycle influence on weather related phenomena.[11]

Pasichnyk explains that the stream of particles flowing from the Sun ionizes Earth's atmosphere, causing a partial vacuum that affects air pressure. As we all know from the nightly weather forecast, variations in air pressure have direct real-world consequences: high-pressure systems tend to bring fair weather and low-pressure systems tend to bring clouds and precipitation. Charged particles emitted from the Sun enter primarily at the poles (as noted earlier), where the planet's magnetic field is weakest. After a lag time of several days, the vacuum created by the ionizing particles causes air to flow from the uppermost (stratospheric) layer of the atmosphere into the layer below (troposphere). The resulting disturbances in air pressure then spread from the polar regions where the particles entered, on down throughout the global system in the form of weather fronts, indicated by capital *L*'s and *H*'s on the weather map.

The more particles that flow, the greater the size and intensity of storms. It is easy to grasp how this might work when one considers that storms are essentially vortices in the atmosphere, energized from time to time by electromagnetic inputs from the Sun. Shortly after being hit by a solar blast, Earth experiences an increase in the frequency of lightning and thunderstorms; this holds true primarily in higher latitudes rather than in equatorial regions, which are largely immune from the polar dynamic. Pasichnyk claims that depending on whether it is a solar minimum or a solar maximum, global thunderstorm activity increases by 50 percent to 70 percent four days after major flares.[12]

So when the Sun shocks Earth's system, lightning and thunder result. And when the Sun shocks Earth's system abnormally, excessive lightning and thunder result. Bullshit? Not according to Vladimir I. Vernadsky, the legendary Russian life scientist (1863–1945) who first proposed the concept of the biosphere, that region of the Earth's surface and atmosphere where life-forms typically dwell. Vernadsky observed that, upon entering Earth's atmosphere,

> Radiant energy [from the Sun] is transformed, on the one hand, into various magnetic and electrical effects; and, on the other, into remarkable chemical, molecular, and atomic processes. We observe these in the form of aurora borealis, lightning, zodiacal light, the luminosity that provides the principal illumination of the sky on dark nights, luminous clouds, and other upper-atmospheric phenomena. This mysterious world of radioactive, electric, magnetic, chemical, and spectroscopic phenomena is constantly moving and is unimaginably diverse.[13]

Pasichnyk theorizes that our weather and climate patterns are ultimately determined by movements in the solar system—specifically, the complex interplay of three vast magnetic fields: our planet's, the Sun's, and what is known as the interplanetary magnetic field (IMF), carried by the solar wind from the Sun to all the planets. (Earth's magnetic field is independent of the one carried to it by the solar wind, instead deriving from the torque of our planet's molten metallic core.) To simplify greatly, imagine three magnets of varying size and shape entering and exiting each other's fields. Two of these magnets, the Sun's and Earth's, rotate on their axes, creating a dynamo effect that generates electrical current. The third magnet is less dynamic and has softer boundaries than its two bolder, whirling counterparts. Depending on the relative position of the three moving magnets, sometimes their forces are additive, other times they balance each other out, and still other times they create curious anomalies, even toss off sparks, in each other's domains. Border crossings among fields create ripple effects that ultimately have consequences for our weather here on Earth. For example, sometimes our planet's own magnetic field crosses the invisible energy barrier between itself and the IMF. Intrusion on this border is known as a sector boundary crossing (SBC). The SBC action creates an energetic ripple effect that is frequently followed by a cyclone, a form of low-pressure system, on Earth. SBCs have been statistically correlated with lightning and thunderstorm frequencies in the Arctic, in the Antarctic, and in midlatitude high-altitude regions. Whether or not this is truly a cause-and-effect relationship remains open to debate, although Pasichnyk certainly believes that the interplay of the magnetic fields has direct impact on our climate, and even on the frequency and intensity of earthquakes and volcanoes.

BALANCING THE BLAME

The theory of how movements in the solar system ultimately shape our climate and seismic environment will undoubtedly be refined as our knowledge is supplemented by the armada of IHY-related research spacecraft poking into the Sun–Earth relationship. Yet it has already become crystal clear that our planet is in a complex and changeable energetic relationship with the Sun and its domain. How the push-pull mechanisms work will be debated and examined for centuries to come. Some effects are cyclical and therefore predictable; others are random and chaotic. The solar system, it would seem, is every bit as moody as its namesake, Sol.

So what strange brew of solar machinations and greenhouse gases led us to our current, overheated state, a.k.a. the Current

Warm Period? Who is more to blame, Sol or soccer moms driving SUVs?

Douglas V. Hoyt and Kenneth H. Schatten offer a compromise:

At the end of the 1700's and in the early years of the 1800's (the "Modern" or "Dalton" Minimum), solar activity dipped, and this era also grew cold. The twentieth century has been marked by gradually increasing levels of solar activity. Cycle 19, peaking in 1958, produced the highest levels of sunspot activity recorded since Galileo's telescopic observations in 1610. The 1990 peak appears to have been the second highest. This global temperature increase approximately parallels solar activity. Recent pulses of greenhouse gases such as carbon dioxide have also caused a warming, so it is not clear how much of the warming can be attributed to each mechanism.[14]

Hoyt and Schatten, whose scholarship is certainly more mainstream than Pasichnyk's, arrive at an evenhanded conclusion concerning the role of sunspots and greenhouse gases in current climate change. Their eminently reasonable position is echoed by Kenneth Lang: "The land-surface temperatures have been correlated with the length of the solar cycle. The yearly mean air capture over land in the northern hemisphere has moved higher or lower, by about 0.2°C, in close synchronism with the solar cycle length during the past 130 years. . . . Short cycles are characteristic of greater solar activity that apparently warms our planet, while longer cycles signify decreased activity on the Sun and cooler times at the Earth's surface," writes Lang—who notes that global sea surface temperatures have also swung up and down with solar cycles over the same 130-year span. "The continued accelerated burning of fossil fuels will someday cause great damage to the environment, so both the developing and industrial nations should do more to stop it. The Sun's activity can nevertheless substantially enhance or moderate this warming."[15]

The stakes in this debate are high. All of the climate ills ascribed to global warming—burgeoning megastorms à la Katrina, advancing desertification such as is baking the Southwest and causing ruinous drought and famine in sub-Saharan Africa, the melting of the poles and the dangerous rise in sea levels, expanding disease vectors as infectious insects range farther afield from their traditional tropical lairs—can now be blamed, at least in part, on the hyperactive Sun. This does not in any way excuse us from controlling the greenhouse gas emissions that still bear the lion's share of responsibility for this slow-motion cataclysm in the making. Quite the contrary, emissions must be reduced more drastically than heretofore considered to achieve the same result. It's simple arithmetic. If the Sun were not a factor, and air pollution accounted for 100 percent of global warming, then a one-third reduction in temperature increase could be achieved by a corresponding one-third reduction in gas emissions. If, on the other hand, only two-thirds of the global warming temperature increase were due to greenhouse gases, then we would have to cut emissions by half—in other words, *more*—in order to reduce warming by a third. (Half of two-thirds is one-third.) From any pragmatic policy perspective, including solar activity in the climate change calculation makes it all the more imperative that we cut greenhouse gases whenever and wherever we can.

Too bad, then, that Gore/IPCC act as though any backtracking from their position that climate change is caused by greenhouse gas emissions would somehow besmirch Science with a capital S. The bluenoses seem to think that centuries of discoveries would be retroactively invalidated if they were shown to be anything less than infallible—and that future scientific progress would forever be retarded by their unforgivable mistake. It's as though they had a moral obligation to defend their position at all costs; the ends justify the means. If this entails derisively dismissing the role the Sun plays in our climate as negligible, and labeling those who disagree as bullshitters, well, those are minor offenses that the climate gods will forgive. But distorting actual data to obscure the Sun's role in global warming, now that's a mortal sin.

7

The Global Cooling Scandal

The scandal known as Climategate broke in November 2009, when thousands of e-mails dating from 1999 to 2009 were hacked from the Climatic Research Unit (CRU) of the University of East Anglia in Britain. Included among these purloined letters were a number of messages from climate change scientists about how to conceal, massage, or spin findings based on the CRU's global temperature record. The offending information indicated that the world's temperatures had actually fallen slightly over the preceding few years. In addition, the CRU saw fit to delete gigabytes of climate data from its computers, presumably to keep the information from being examined by independent authorities. Fortunately, most of that data was backed up on other servers, though access to it still remains an issue. The timing of the hack attack on CRU was certainly suspicious, coming just in time to damage the United Nations' ill-fated Global Climate Conference, attended by President Obama and many other world leaders, held in Copenhagen the following month, December 2009. The suspicion of foul play by naysayer partisans was reinforced when a second wave of East Anglia e-mails was released in November 2011, weeks before the UN climate conference in Durban, South Africa.

At first I figured Climategate was just a case of some Podunk U cooking their results to kiss bluenose butt. But, as I soon discovered, East Anglia is no fringe institution in the climate change debate; its CRU provides the hard temperature data upon which the IPCC bases its projections and its trillion-dollar policy recommendations. Most of the purloined e-mails were neutral or ambiguous, interpretation depending largely on one's ideological slant. Some messages included complaints of data being twisted to fit foregone conclusions; others alluded to the suppression of studies that did not toe the party line. But this one, from Phillip Jones, head of East Anglia's CRU, is truly damning: "I've just completed Mike's *Nature* [the science journal] trick of adding in the real temps to each series for the last 20 years [that is, from 1981 onward] and from 1961 for Keith's to hide the decline."[1]

Jones now claims that the word "trick," in this context, refers not to subterfuge but rather to a handy statistical method used "to bring two or more different kinds of data sets together in a legitimate fashion by a technique that has been reviewed by a broad array of experts in the field."[2] Maybe, but no matter how you slice it, it's still baloney, as they say. Any manipulation designed to "hide the decline" in temperatures is falsification, pure and simple. Clive Crook, writing for the *Atlantic*, nails it: "Well.

It seems to me, and I daresay to other open-minded readers, that the talk in the e-mails of a 'trick to hide the decline' raised the reasonable suspicion that a trick had been used to hide the decline."[3]

Also tarred in the ruckus was Michael Mann, of "hockey stick graph" fame. Mann, now at Pennsylvania State University, had corresponded steadily with Jones, the two of them commiserating on how to deal with those pesky naysayers. But let's give the blue-noses the benefit of the doubt for a moment, since they do their work in the name of the very worthy cause of saving civilization from overheating, mass death, and destruction. Are Gore/IPCC justified in peremptorily dismissing opposing scientific arguments on the grounds that these are not normal circumstances? Is the climate change state of emergency dire enough to shut out the naysayers, even though under normal circumstances their perspectives might be fodder for constructive intellectual debate? Have we reached the point of desperation, where anything that weakens our collective resolve to curb the emission of greenhouse gases, which are clearly culpable in the current global warming, endangers us unacceptably? Has the mission of the scientists who discovered this looming threat morphed from presenting the results objectively to championing policy implications thereof by any means necessary?

"Ironically, the only way to really find out if phenomena like sunspots and solar wind are playing a larger role in climate change than most scientists now believe would be to significantly reduce our carbon emissions. Only in the absence of that potential driver will researchers be able to tell for sure how much impact natural influences have on our climate," observes a recent *Scientific American* editorial.[4]

It sure is easy for me up on my high horse to condemn the bluenoses, and just as easy to imagine them condemning me. "So what if the Sun *does* have something to do with this crisis of runaway global warming? Unless you've got a way to calm that star down, quit interrupting! It's hard enough to get people even to consider breaking their greenhouse gas habits, much less undergo all the sacrifices necessary—and they *are* sacrifices, try as we might to depict them as innovative opportunities for whatever—without you feeding them escapist fantasies about sunspots. And it's 10 times harder today than it was even a few years ago, with the collapsing world economy." Assessing my behavior, Chairman Mao might have been moved to repeat his famous remark, "You are a reckless adventurer whose actions play into the hands of the existing order."[5]

There is no denying the broader IPCC position that trivial temperature fluctuations over the course of a few years do not in and of themselves disprove a climatic trend 150 years in the making. And it is easy to see how the bluenoses would see this minor but untimely decline as a real momentum killer and therefore feel that they could, in good conscience, justify covering it all up. It's kind of like those freak blizzards that seemed to hit Washington, D.C., every time President Obama talked about global warming—one-off occurrences that prove nothing, at least not scientifically, but that nonetheless strengthen the naysayers' hand. Most mainstream news cut the bluenoses a lot of slack on the Climategate story, referring to the data anomalies as minor, some even focusing more on the illegality of how the information was acquired than on the substance of what was revealed. (When reporters scream about how wrong it is to mine a news source, the whiff of bias is ripe.) It turns out that a BBC weatherman had in fact received copies of the damning East Anglia e-mails a month before the scandal broke and reported nothing. No coercion there. Notable exceptions to this journalistic timidity to take on the climate change establishment were right-wing outlets such as Fox News in the United States and the *Telegraph* in Great Britain, which covered the scandal with their usual gusto, as did right-wing talk radio.

COLANINNO MINIMUM: THE SMOKING GUN

What got lost in all the Climategate hubbub is the smoking gun. The drop in temperatures which the CRU was apparently trying to cover up coincided closely with a drop in the number of sunspots on the Sun's face. Beginning in 2007, the Sun lapsed into a profound lethargy, setting record lows for energy output. This lull is known as the Colaninno Minimum, named after Robin Colaninno, who first predicted it in 2006. Astonishingly, the years 2007, 2008, and 2009 all rank among the bottom 10 of the past 100 in terms of solar activity!

Solar physicists now believe that the Sun's conveyor belt system got snarled up, causing the sunspots to vanish. Tony Phillips, NASA's editor-in-chief, offers this explanation:

> A vast system of plasma currents called "meridional flows" (like ocean currents on Earth) travel along the sun's surface, plunge inward around the poles, and pop up again near the sun's equator. These looping currents play a key role in the 11-year solar cycle. When sunspots begin to decay, surface currents sweep up their magnetic remains and pull them down inside the star; 300,000 km

**below the surface, the sun's dynamo amplifies the decay-
ing magnetic fields. Re-animated sunspots become buoy-
ant and bob up to the surface like a cork in water—voila! A
new solar cycle is born.[6]**

The minimum that began in 2007 was caused by the speedup
of the solar conveyor belt. This interfered with the sunspot re-
magnetization process, preventing sunspots from bobbing back
up to the surface as they normally would. These neglected spots
soon dissipated. Even though the conveyor belt returned to
normal speed in 2008 or early 2009, scientists believe, there were
at that point no more sunspots to remagnetize, so the minimum
continued unrelieved. Though we tend to think more about
solar maxima and the storm potential these interludes represent,
the fact of the matter is that these peaks rarely last for very long,
and are usually over after a couple big storms. Solar minima, by
contrast, can go on indefinitely, such as the three-year Colaninno
disappearance, 2007–2009, or, more famously, the 70-year stretch
known as the Maunder Minimum. Thus the *lack* of sunspots can
and often does end up having a greater impact on our climate
then a *surfeit* thereof.

So sunspot activity dipped, and so did temperatures on Earth.
Yet another exhibit for the Moody Sun Hypothesis defense. Then
there's the fact that Earth's thermosphere, a layer of the upper
atmosphere ranging in altitude from 90 kilometers to 600-plus
kilometers, collapsed during the Colaninno Minimum. Phillips
describes that phenomenon:

**It [the thermosphere] is a realm of meteors, auroras and
satellites, which skim through the thermosphere as they**

circle Earth. It is also where solar radiation makes first contact with our planet. The thermosphere intercepts extreme ultraviolet (EUV) photons from the sun before they can reach the ground. When solar activity is high, solar EUV warms the thermosphere, causing it to puff up like a marshmallow held over a camp fire. (This heating can raise temperatures as high as 1400K—hence the name *thermosphere*.) When solar activity is low, the opposite happens.[7]

Researchers say that the Colaninno collapse was the most severe contraction of the thermosphere in over 40 years. But how do they know this? Was there a thud? Scientists have been measuring the decay rate of more than 5,000 satellites since 1967—in other words, for all but the opening decade of the Space Age. The hotter the thermosphere, the worse the aerodynamic drag on the satellites, the more rapid the decay. So for spacecraft and their handlers, the sunspot disappearance of 2007–2009 was a respite from the usual damaging friction. A dream vacation, one might go so far as to say, since it turns out that the Colaninno collapse was two to three times greater than the drop-off in the Sun's activity alone could explain.

What else could be affecting the thermosphere so dramatically as to help engender its collapse? Phillips speculates that climate trends in the troposphere (the lowest portion of the atmosphere, in which we live) may somehow have contributed to the upper atmosphere's cooling deflation. "One thing is clear," he writes. "During long minima, strange things happen. In 2008–2009, the sun's global magnetic field weakened and the solar wind subsided. Cosmic rays normally held at bay by the sun's windy magnetism surged into the inner solar system."[8]

As we have seen, cosmic rays make clouds, which cool things off. During the Colaninno Minimum, the deepest solar minimum in a century, there were lots more cosmic rays around than normal, 19 percent more than anything recorded in the previous 50 years, according to Richard Mewaldt of the California Institute of Technology.[9] Cosmic rays are subatomic particles accelerated to near light speed by the shock waves of deep space explosions. When they crash into Earth's upper atmosphere, the impact creates a host of secondary particles, which rain down into the lower atmosphere, eventually serving as nuclei around which moisture coalesces. It's like a cloud being seeded, but on an ultra-microscopic scale. This, in a nutshell, is how cosmic rays stimulate cloud production in the lower atmosphere—those clouds that cool us by shading out the Sun and sending down the rain. So when the Sun is in a quiet mood, not shouldering out the rest of the universe the way it usually does, we on Earth cool off a bit as well. Here we have yet another example supporting the Moody Sun Hypothesis that vagaries in the Sun's behavior end up affecting our climate and our lives in ways we never imagined.

TIME FOR WORST-CASE SCENARIOS?

That is why the East Anglia Climategate scandal is so insidious. It's not just about the planet's temperature ticking down a few basis points. No one at IPCC ever claimed that new heat records would be set each and every year. Rather, the key fact being obscured with all the data fudging is that *the cooling of Earth directly coincided with the cooling of the Sun.*

It is only natural that any body, including a heavenly one, might need to rest after so many decades of booming. Was the

Sun just taking a catnap with this recent Colaninno lull, or will it go into deep hibernation once the 2012–2013 solar maximum passes? We could be in for some good news if we are indeed entering a new minimum, because more cosmic rays would get through and the resulting cloud cover just might help offset global warming. Not sufficiently to neutralize the more powerful warming influence of greenhouse gases, unfortunately—but, knock wood, just enough to give us a little extra time to implement the Gore/IPCC climate change agenda. A sunspot lull is something worth hoping, praying, and rooting for.

We can knock all the wood we want that the Sun will calm down and help mitigate the warming that is either completely out of our control or in the hands of a Higher Power whom we might beseech. But what if we do not plunge into an extended solar minimum? What if there is no reprieve and global warming cannot be controlled? Acknowledging that climate change is cyclical means acknowledging its inevitability, means acknowledging that we should at least consider investing as much of our time, effort, and treasure in preparing for a warmer world as we now do in trying to stave off such an eventuality. If global warming is a given, impossible to stop because it is natural to the climate's ebb and flow and/or because greenhouse gas emissions will never be curbed in time, what defensive measures must we take to minimize loss of life and property? Sell the beach house? Dump the solar energy stock and instead invest in air-conditioning companies? Horde water?

Accepting that global warming is our inevitable fate is kind of like declaring bankruptcy—a humiliating and problematic thing to do, though sometimes it's the only cure for the problem. It's a solution that no one will champion because it's a craven admission of defeat. Which creditors will get stiffed the worst is hardly an inspiring conversation, but you damn well want to have a seat at the table when the assets start being sold off and divvied up. Should water desalinization plants be built at either end of the Sahel region, the parched ecological corridor that runs across Africa (separating the Sahara to the north and grasslands to the south), to serve the teeming millions who will be hardest hit by the drought? Or should we start with Baja, a convenient drive from sunny San Diego? If there's funding to construct artificial barrier reefs along the coasts to control flooding, do we protect Bangladesh or the Hamptons? Proactively relocating hordes of eco-refugees from drought-blighted areas might be a good idea, but which nations get the, uh, honor of embracing these trauma-tized aliens as their own? We have to start thinking of worst-case fallback scenarios, if for no other reason than that these scenarios might light the fire under us that all the "stop global warming" rhetoric has thus far failed to do.

Clearly we have reached a dialectical moment in the environ-mental debate. A new thesis must be forged. Bluenoses and nay-sayers must come together and agree that current climate change is being caused by two converging forces: 1) the accumulation of man-made greenhouse gases and 2) recent fluctuations in solar activity. The Sun we cannot do a thing about, except to monitor its behavior as closely as possible. The emission and accumula-tion of man-made greenhouse gases aggravating our climate and endangering our way of life we can, perhaps, control. Facing the problem honestly rather than dogmatically is the only way for us to unite in the common purpose of saving our beloved civiliza-tion. And that's no bullshit.

THE SUNBIRDS RETURN

Omens are what you make of them. The fact that four ancient golden sunbirds representing north, south, east, and west recently returned to China may or may not be more than coincidental to the resurgence of our fascination with the Sun and its fickle mastery over our climate. In February 2001, an astounding archaeological discovery was made of a 3,000-year-old Sun-worshipping culture, known as the Shu, centered in or near what is now known as Jinsha, a suburb of Chengdu, capital city of the Sichuan province of western China. Among the hundreds of treasures on display in the Jinsha Site Museum, the supreme artifact is made of gold: four sunbirds circling the Sun, which blazes with 12 spiral rays, a veritable diadem penetrating our minds with light and reason. What a revelation! This exquisite image has, since its discovery, been formally adopted by the Chinese Ministry of Culture as the official symbol of China's cultural heritage. It was even launched into space on the Shenzhou 6 manned spacecraft in 2005, to illuminate the void.

The sunbirds are descended from the legend of Hou Yih, the hero of Chinese mythology who saved Earth from being burned out by ten blazing suns, a metaphor for global warming if there ever was one. Atop a giant mulberry tree towering in the eastern sky, there lived in ancient China ten golden sunbirds, children of the god and goddess of the sky. At the start of each day, the Sky goddess would descend from the heavens in her dragon-drawn chariot and stop off at the mulberry tree to collect one of the sunbirds, who would spend the day working as the Sun, riding across the sky in the chariot and shedding heat and sunshine down onto Earth below. Day after day, year after year, this ritual was repeated, with each sunbird working one out of ten days, and playing with siblings the other nine. Although the arrangement was close to ideal, the sunbirds grew bored; they chafed to try something new. One day, they decided to fly up together, ten sunbirds all at once dominating the sky. Quickly Earth below grew intolerably hot. Water evaporated, crops withered, peasants died of thirst, monsters ravaged the landscape. When the people's prayers for relief from the unbearable heat reached the Sky god, he became very angry at his sunbirds for flitting recklessly about in the sky. He summoned Hou Yih, the god of archery, and commanded him to point his bow in the air to scare the sunbirds into obedience. But the sunbirds paid Hou Yih no heed and refused his command to return to their tree.

The emperor of China then commanded Hou Yih to shoot an arrow skyward. It pierced one of the sunbirds, which crashed to the ground in flames. He shot a second arrow, then a third, fourth, fifth . . . until nine out of ten sunbirds had been felled. With each death, Earth cooled a bit; clouds, mist, and dew returned; springs and rivers bubbled up; grass, trees, and crops flourished anew. The archer drew his bow for one last shot, but the emperor ordered him not to extinguish the lone remaining sunbird. So it was that Hou Yih spared one final sunbird, which was condemned to trudge across the sky in solitude for eternity, bringing (a moderate amount of) light and warmth to the world.

Was there a time way back when that folks actually believed this story literally? Didn't common sense say pretty much the same thing 3,000 years ago as it does today, that the Sun isn't a bunch of birds that can be shot out of the sky by arrows? Or was there some other metaphorical thing going on back then? Perhaps it is analogous to the question of how many Roman Catholics actually believe that they are eating the physical body and drinking the actual blood of Jesus Christ upon receiving Communion. My feeling is that extreme doctrinal beliefs are rarely held literally, yet the fact that one is *supposed* to believe in such miracles somehow solemnizes the whole worshipping process. But what we modern skeptics, severely disinclined toward fairy-tales, must always remember is that just because a story doesn't make *literal* sense doesn't mean it is devoid of value or truth. The tale of Hou Yih and the golden sunbirds elegantly captures the essence of our current conundrum: a hyperactive Sun is causing a climate problem, and action from the ground is what will solve it. Yes, this ancient myth overstates the role of solar variability; and no, it does not specify that the needed ground action concerns the emission of greenhouse gases. Bull's-eye predictions are hard to come by 3,000 years in advance. But if nothing else, isn't it a helpful coincidence that a smashing archaeological discovery with a timely solar parable has emerged from obscurity just as we turn our attention to the Sun's impact on our climate?

Caveat to any would-be Sun-shooters today: the Emperor was grateful to Hou Yih for saving the world and hailed him as a hero. But the Sky god was furious that nine of his children had been killed in the process and thus took revenge by stripping Hou Yih of his immortality. The archer wandered Earth searching for a way to reclaim everlasting life. When he climbed a mountain range and entreated the Queen of the West for help, she commanded him to build her a palace of jade. After he finished, she rewarded Hou Yih with an elixir that would allow him to live forever. She instructed him to fast and meditate for one year before taking the elixir in order to regain the wisdom necessary to live eternally. Otherwise, misery would be his forever. Hou Yih dutifully set to his task. However, his wife, Chang Oh, spied the elixir and swallowed it all. The Queen of the West was outraged by the theft and banished Chang Oh to the Moon. Hou Yih leaped up to rescue his wife as she soared heavenward, but he was too late. Soon she disappeared inside the Moon, which lit up radiantly. This, according to the folktale, is how the Moon got its glow. At the death of his wife, Hou Yih grew bitter and tyrannical, eventually becoming so obnoxious that an angry mob beat him to death. Nonetheless, his spirit ascended to the Sun. Thus it was that Hou Yih and Chang Oh came to embody the yang and the yin, the Sun and the Moon.

8

Twenty-First-Century Sun Worship

Will anthropologists of the future look back and conclude that we were a Sun-worshipping civilization that idolized David Hasselhoff and Pamela Anderson? They would be kind of right, you know. *Baywatch* is among the most popular television programs in history. The L.A. lifeguard, Sun-and-fun, beach bimbo/bozo show sold its lifestyle and values, such as they were, in almost as many as countries as Coca-Cola sells its syrup. What a powerhouse! The comely Anderson happened to wear some scrunchy, soft boots while shooting the show, and boom, Uggs became a global phenomenon. If the show's 1.1 billion weekly viewers had prostrated themselves shoulder to shoulder, they could have spanned the diameter of the Sun.

Baywatch went off the air in 2001. If its producers are planning a comeback, they might scout a new look for the show at Malibu

Makos Surf Camp in Malibu, California, where beach meets Halloween. Every morning before the campers wet-suit up to surf and boogie-board, their faces are slathered in zinc oxide from forehead to chin. Boys squirm as they get white-faced by their parents. Girls generally present themselves genially to be made up in hot pink, yellow, and/or green. By the end of the week, with their eyes bloodshot and their hair almost white from too much salt and Sun, and various layers of SPF (sun protection factor) products staining their skin—those zinc oxide ointments are formulated not to wash off in water—the kids look like Sun-bleached refugees from some aboriginal dreamtime, or like Munchkin extras from *The Wizard of Oz*.

Whatever happened to "It's better to look good than to feel good," that modern-day mantra? Skin cancer is what happened. Skin cancer is a lifestyle disease, born of too much leisure and fitness, too many bikinis and sculpted pecs, and too much desire to show the whole package off in the blazing Sun. Also known as skin carcinoma, this group of sometimes fatal diseases is the very high price some of us are forced to pay for soaking up more rays than our skin can naturally handle. Coco Chanel started the epidemic in the 1920s by getting the first "fashionable tan," lying out on the deck of the Duke of Westminster's yacht so she could turn golden brown like all the hunky deckhands instead of staying pasty and translucent like the English aristocrats cowering in their cabins. It is often remarked that people like to get a tan because it makes them look healthy, visual proof positive that they have been outdoors, active in the fresh air and sunshine. Furthermore, a tan tends to make one look *thinner*, which in today's pinched cultural ethos is even more of an asset than looking healthy. Just as dark clothes are generally more slimming to the appearance than light garments of the same cut, a darkening of the skin has a flattering effect.

Can't blame Coco for the mass migration to the sunbelt, though. That falls on the head of one Willis Haviland Carrier, founder of the Carrier Air Conditioning Company of America, who in 1902 invented the earliest mass-producible version of the machine that would go on to make sunbelt summers livable. The sunbelt migration has increased the prevalence of skin cancer by increasing the numbers of fair-skinned people living in regions of bright sunlight, such as Florida, Arizona, and California, Spain, Portugal, and southern France, Dubai and the Arab Emirates. Ironically, the spread of the same air-conditioning technology that enabled such migration has compounded the skin cancer epidemic by hastening the depletion of stratospheric ozone. As discussed in chapter 6, leakage of Freon and other chlorofluo-rocarbon refrigerants from compressors into the atmosphere has thinned the ozone layer, which in turn has caused skin cancer rates to rise. An increase in skin cancer has also been noted in areas such as the United Kingdom and Quebec, nearer the poles (where ozone holes typically occur), and where Caucasian inhab-itants tend to be paler and more susceptible to sunburn.

Like moths to the flame, yet also like subterranean prisoners moving toward the light, migrants around the world are choos-ing to escape the cold, dark winter permanently. It is said that the Rose Bowl Parade, normally a sunny, sparkling event held each year on New Year's Day in Pasadena, California, has done more to spur migration to Southern California than anything else. Shiverers in the Northeast and Midwest decide right then and there, watching their TV, to warm up once and for all. It's as though anyone who can manage it makes the move. Rather than

"Why live in the sunshine?" the more pressing question is "Why *not* live in the sunshine?" No one stops to wonder, "What if I get skin cancer?"

Surprisingly, no one really knows how many cases of skin cancer there are each year in the United States. This is because the less serious and far more common forms of the disease (see below), such as the type a dermatologist might burn off during a routine office visit, are normally not reported or included in health statistics. Anecdotally, these comparatively innocuous skin cancers are much more prevalent in sunbelt states, though few hard figures are available. According to the National Cancer Institute, more than 2 million Americans are treated annually for some form of the disease by the time they reach 65, although it is unclear how this statistic was derived.[1] Whatever the exact numbers, skin cancer cannot be dismissed as a low-percentage possibility. It is a statistically reasonable worry. The good news is that the Sun does indeed lose most of the time, in the sense of not causing permanent physical harm to the people and animals upon which it inflicts its damaging rays. More than 90 percent of all skin cancer cases are resolved safely, though not necessarily completely. The disease can disfigure.

Driving up the Pacific Coast Highway north of Santa Monica, California, one occasionally passes signs indicating the level of fire danger for that day. Cloudy damp days are rated *Low*, and sunny days, particularly when the hot, dry winds blow in from deserts to the east, are rated *High*. Why not another set of signs rating the UV threat for the day? As a parent, it sure would be helpful to have a big red flashing *10* to point to when trying to get the kids to stay inside, or at least to slather on the sunscreen and cover up carefully. If there is no government funding available, maybe Coppertone could sponsor this program. The basic information is there for the asking. The U.S. National Weather Service began publishing a Sunburn Index in 1994, updated and renamed the UV Index in 2004 to incorporate guidelines established by the World Health Organization (WHO) in its Global Solar UV Index, by far the most sophisticated tracker of the Sun's moods as they vary from day to day. WHO's index is meant to indicate the maximum threat possible on any given day, which (all other things being equal) comes in the early afternoon. It takes into account the season, latitude, and cloud cover. Basically, the scale goes from 1 to 20. Anything over 6 is considered risky, and anything over 10 means you should stay inside. Clearly, though, the relevance of those assessments varies depending on the skin tone and SPF-preparedness of the individual.

How much of surrendering one's body to Sun worship on the beach is ritual, and how much is recreation? Forgoing a familiar pleasure is much harder than abstaining from it in the first place, and harder still when said pleasure does not feel like a vice but rather like a healthy, natural activity, such as is the case with lying out at the beach, body caressed by sunshine and sea breeze and refreshed by waves. Giving all this up because it might be bad for you seems counterintuitive, absurd. Funny to think, though, that it has been less than a century since Coco lay out on the deck of the duke's yacht. Despite its primal feel, sunbathing as a popular recreational activity really just started yesterday, in terms of our million-year human history. And what an industry near nakedness has spawned! Resorts and public beaches now line most coasts. True, the threat of sunburn and skin cancer has dampened our enthusiasm a bit. Now sunblock sales soar, and more hats, parasols, and other protective garments dot the summerscape each year.

It never crosses our minds that the sunlight shining down on us might be fluctuating in spectral composition, that the mixture of ultraviolet and infrared rays might vary, with corresponding health impacts on bodies, ecosystems, and the global climate as a whole. But according to data from research satellites monitoring the Sun, the composition of sunlight does indeed vary over the course of the solar cycle, the aforementioned 11-year period over which the Sun's activity typically rises and falls. One of the most surprising findings is that solar output varies a lot less in the middle part of the spectrum than at the extremes. As the Sun nears its maximum, a 0.1 percent increase in overall solar output might typically be accompanied by a 6.0 percent jump in ultraviolet (UV) radiation. Such UV spikes can have profound effects on our physical well-being, and are closely correlated with heightened rates of sunburn, skin cancer, cataracts, and other eye disorders.[2]

The Moody Sun Hypothesis suggests that the likelihood of your getting a sunburn, skin cancer, or eye disorder might well depend not only on what season but also on what year—at what point in the 11-year solar cycle—you do your sunbathing. Therefore, we should all seriously consider limiting our Sun exposure during solar climax summers. The problem is that very few of us think in terms of such large chunks of time. The longest cycle most of us are used to is four years, for presidential elections, the World Cup in soccer, and the Olympics, winter or summer. The federal government could undertake to remind us to be especially careful during solar climax years, though resort industry lobbyists would cry foul and blame whichever incumbents supported the warnings for trying to push their industry into recession. At this writing, the next climax summer will come in 2013, and I do not look forward to being the only parent telling his kids that they cannot go to surf camp because the sunspots are acting up. What's the tradeoff? My kids, like most of the Malibu Makos, are pretty fair-skinned. And all it seems to take for UV radiation to trigger a future malignancy is for a very small number of cells to have their DNA machinery corrupted, at which point they start reproducing uncontrollably. Are the pleasures of spending the summer at the beach worth the risk? Or should I just have the kids laminated?

THE SUN-SKIN DART GAME

To get how sunshine causes skin cancer, imagine that the Sun is throwing a zillion darts of UV radiation in your direction every second. About 95 percent of the darts are in the UVA range, with wavelengths from 320 to 400 nanometers. (A nanometer is a billionth of a meter.) Virtually all of the remainder that make the 93-million-mile voyage all the way from the Sun to you are UVB darts, with wavelengths in the 260 to 320 nanometer range. (UVC darts, with wavelengths ranging down to 100 nanometers, are deflected en route.) Only a tiny cohort of UVA and UVB darts get past all the obstacles presented by clouds, particles in the air, and shade cover. The preponderant majority of darts with your name on them are stopped by your clothes or your skin. But every now and then a dart is thrown with such force and trajectory that it gets all the way through the epidermis, the outermost layer of your skin, down to the base level, where the small, round basal cells are found. A pretty good shot, though if this were indeed a game of darts, the Sun would score only 10 points, compared to 50 for a bull's-eye. Basal cells carpet the whole bottom layer of the epidermis, so the odds of the Sun eventually hitting at least a few of them are not so prohibitive.

If the tip of the UV dart hits the basal cell's DNA package, that cell may become cancerous. Basal cell cancers, or carcinomas, are by far the most common, accounting for eight or nine out of ten skin cancer cases. Eighty percent of these "sores that don't heal" form on the head and the neck; virtually all of them are readily visible, usually as red, raised, scaly patches. (Precancerous scaly patches known as actinic keratosis form in a similar manner.) Basal carcinomas are slow-growing and rarely metastasize, though they can be disfiguring. They are the least dangerous form of skin cancer, with a survival rate approaching 100 percent; most survivors enjoy full medical and cosmetic recovery.

To score, say, 25 points, the Sun's got a trickier shot to make. Instead of driving all the way down to the base of the skin, the UV dart has got to prick the skin's surface, which is made up of dead skin cells, and lodge in the layer of living cells just below. Skin cancers that originate in this layer of squamous cells—cells flat and shaped like fish scales—are usually more serious than basal cell cancers, and they can spread, especially if they develop on mucous membranes such as a lip or inside a nose. However, if caught early on, as fortunately often happens (since these lesions also occur in plain sight on the surface of the skin), squamous carcinoma recovery rates approach 100 percent, though permanent disfigurement sometimes results.

The 50-point shot is a truly sick victory, since it hits a melanocyte, the very type of pigment-producing cell charged with protecting our skin from the Sun. The more exposure to sunlight, the more melanin (or pigment) the body produces—and thus the darker (tanner) and more capable of absorbing light rays the skin becomes, although the efficiency of the tanning process varies widely from individual to individual. Melanoma is caused by the impact of a UVB dart into a melanocyte's DNA machinery. Unlike basal cells or squamous cells, melanocytes are quite mobile, meaning that they can spread the cancer they bear with deadly effectiveness. Melanoma is identified by brown, irregular patches on the skin. If it is treated in time, recovery is possible. Unlike the other forms of skin cancer, however, it is frequently fatal; in fact, in the Western Hemisphere, melanoma is among the deadliest of cancers. Curiously, the U.S. states with the highest incidence of melanoma are located in the northwestern and northeastern regions, hardly the sunbathing Meccas of the country. This is according to the Environmental Protection Agency, which tracked the disease from 2003 to 2007.[3] In all likelihood, this phenomenon is related to the fact that stratospheric ozone depletion is severest near the poles.

WORSHIP AT YOUR OWN RISK

Were Sol an observer, he would no doubt take pride in the tribute we render him, even at our own peril. I confess to once having been swept away by Sun worship, and I pity the current majority who, ever mindful of the dangers of skin cancer, deny themselves such religious ecstasy. Eons before becoming a commendable father obsessed with sunscreening the tips of his children's ears, I was 22, a graduate student in comparative literature at the University of California at San Diego. Black's Beach, the nude beach right next to campus, had been daring me for months, so one fine spring day I finally got up the nerve. To get to the beach one had to climb down some pretty steep cliffs, dodging the parade of impossibly fit surfers running barefoot back up the rocky trail, surfboards shouldered high. Upon reaching the top, they would ditch their boards and launch themselves off the cliffs in hang gliders, soaring out over the Pacific until some inner circadian alarm told them it was time to go back down to the beach and spank the sparkling waves. Being from Brooklyn, where we played in fire hydrants, all of this amazed me.

"Sun feeds the muscles," declared Herodotus, who reported in the fifth century BCE that ancient Greek athletes trained naked out of doors specifically for the purpose of exposing all of their muscles to the benefits of sunlight. Genuine athletes look all the nobler when naked in their pursuits; the unity of body, sand, ball, and sunshine is every bit as organic as Earth, air, fire, and water.

Back in San Diego, down at the beach the first thing I beheld was a full-on, co-ed, set-and-spike nude volleyball match in which the all-tans appeared to be beating the white-butts, and not a drop of SPF on any of them. The beauty of that hard-fought nude volleyball match was undeniable. But was it *worship*? There was definitely something sacred about it, a communing not to be violated by the uninitiated. I felt like an intruder, like one of those Christmas/Easter interlopers who partake of Holy Communion even though they suspect the Church forbids them to do so. But no more than 50 yards downshore there was another group, this one profane. To them, nude volleyball was nothing more than a party game in which participants petted, batted, and guzzled. Between the two groups one beheld good and evil, nakedly juxtaposed. I despised this second, decadent bunch, much as I would years later despise the Limelight, a deconsecrated church turned trashy New York discotheque. God would probably agree with both moral judgments, but the Sun god? Would he/she really care one way or the other, or are the concepts of good and evil utterly beside the point to whatever spirit might emanate from a behemoth ball of flame?

I found some answers to this perplexing moral question 30-odd years later when confronted with the ancient Egyptian Sun god, Ra. In 2011, a revolutionary movement in Egypt succeeded in chasing President Hosni Mubarak from the office he had held—some say, held hostage—for almost 30 years. Although at this writing it is still unclear who or what group will next control that nation, there is no doubt that one of the sturdiest bulwarks against the seizure of that country by extremist elements is the greatness of Egypt's pre-Islamic past. Thank Ra and his ancient disciples for doing their jobs so well that the civilization they created has lasted, physically and to some extent culturally, for 5,000 years (or much longer, as some Egyptologists claim). Ra and his retinue continue serving the Egyptian people, who, though they no longer worship the Sun, profit from ancestors who did. Were it not for these ancient pagans and their gods, tourism, Egypt's largest industry in terms of both monetary and cultural wealth, would barely exist. Like Greece, also dependent on its ancients for sustenance, Egypt would not have attained the global prominence it has. Napoleon Bonaparte would never have bothered to invade.

What theology would be associated with Ra were he worshipped today? The principle virtue in any solar religion would seem to be not goodness, but reliability. The Sun is the greatest constant in the history of humanity, greater even than God, because some people do not believe in God. Our star has been our sole provider of warmth and light, rising and setting infallibly for almost five billion years. What's more reliable than the Pyramids? Woody Allen's famous quip, "Ninety percent of life is showing up," might be adapted quite readily into a Ra/Sun commandment.

THOU SHALT SHOW UP

Showing up is what matters. Whether showing up to do good or to do evil would, in a solar-centric religion, seem a second-order distinction.

Ra disappears just as reliably as he appears, dying every night and getting reborn every morning. According to *The Sungod's Journey Through the Netherworld,* by Andreas Schweizer Ra (or "Re," as Schweizer prefers) boards a boat at the end of each day and sets sail for the netherworld.[4] Midway between sunset and sunrise, he mystically unites with the mummified corpse of Osiris, god of the dead. At first only pharaohs had the privilege of becoming one with the Sun god's soul, but eventually the doctrine broadened to provide that all who die in a blessed state could make the same "night-sea" journey and attain everlasting life and light in the netherworld. The same Ra-voyage happens, metaphorically, every night we sleep: in the depths of our slumber, we recharge and reconnect with the eternal energy of creativity. I'm not sure to what extent the ancient Egyptians grasped the notion of Earth's rotation, and therefore the fact that when Ra was supposed to be communing with Osiris's mummy, he was actually sparkling midday on the opposite side of the globe. Should we condescend to "forgive" this inconsistency as commensurate with the limited scientific knowledge of those early times? I prefer to consider it all an object lesson that truly reliable Sun worshippers revere Ra with equal fervor at noon and midnight, even when we know he's half a world away and cannot see us.

Ra may also move in mysterious ways. The idea of consciously worshipping the Sun as some sort of deity, actually getting down on bended knee or, say, prostrating oneself at dawn, noon, midafternoon, sunset, and midnight, does seem strange, though it is really no more irrational than worshipping God. On the one hand, the Sun cannot hear our prayers or be affected by them in any way, although we fragile mortals might gain strength and bounty by aligning ourselves with the source of Earth's blessings, secure in the knowledge that the Sun exists and empowers our lives. By contrast, if God has the anthropomorphic and/or empathetic dimension that most believers hold, he/she is therefore capable of responding to prayer and affecting our outcomes favorably. That is, if such an entity exists. There is no denying that God, no matter how fervently we believe, might turn out to be just a wishful construct. How odd that we give thanks to God, an entity that virtually no one has verifiably seen or heard, rather than to the Sun, which everyone sees and/or feels every day. Worship is a way to show appreciation for life's blessings as well as a way to encourage these blessings to continue or increase. Unlike God, Ra does not concern himself with personal outcomes. It would be less efficacious to implore Ra, for example, to help me land a specific job than it would be to ask that many good job opportunities cross my path. How the personal details work out is none of the Sun god's concern.

George Carlin understands. In "Religion Is Bullshit," his popular comedy sketch now widely available on the Internet, Carlin propounds a hip, hilarious, and amoral cosmology. He denounces the whole notion of God as vengeful and implausible and instead declares that he has taken up worshipping the Sun, which, unlike God, can actually be seen. Providing warmth, light, food, flowers in the park, albeit also the occasional skin cancer, the Sun, Carlin adds, has never made him feel sinful or unworthy. How to show appreciation for the great solar entity? The "bullshit" artist let his shtick do the talking, but most of us have to find an alternate means of self-expression. In *The Pillar of Celestial Fire*, Robert Cox suggests—what else?—meditation, which is something he believes that the Sun has been doing for billions of years. "This ball of celestial fire is so filled with spiritual power that it resembles a shining crystal or radiant diamond. It may be understood as the crystalline heart of the Sun."[5] Go figure. It turns out that the vibrational frequency of the "egg-shaped cocoon of luminous subtle matter that lies at the center of the solar orb," as Cox puts it, exactly matches the rate at which Carlin is spinning in his grave.

Kidding aside, there may be some merit in our trying to relate to the Sun as a sentient being even though there are no real indications that that's what it is. Anthropomorphizing an inanimate entity does tend to sharpen one's focus on the subject in question. Sure, there is a trade-off. But aren't heightened powers of concentration worth obtaining, even at the expense of proceeding from erroneous premises? The human need to humanize is undeniable. Is this what led the ancient Sun god to morph into the shining Lord of Christianity? Perhaps the Sun is most constructively viewed as the gateway to God. As the greatest manifestation of divine power in our lives, it is uniquely worthy of veneration. And at this time of accelerating climate disorder and solar threats to our infrastructure, should we not take special pains to integrate the Sun into our worship? Perhaps God would be better served if we directed our prayers straight to his great solar masterwork rather than launching them uncertainly to fuzzy and shifting images of the deity—him, her, hesh (he/she), it, they, all (or none) of the above? Not that we should pray to Ra, exactly, but much as Christians pray to God in the name of Jesus, to God in the name of his Sun.

9

The Sun Also Heals You

Swimming is my favorite form of exercise and also my vice, considering how much it costs to heat the chilly backyard pool. One evening, just as the Sun was going down, my left shoulder began hurting as I started my workout, so rather than relying on the usual freestyle stroke, I decided to do all of my laps under-water—about 45 minutes' worth. It's a saltwater pool, very little chlorine, so I swam without goggles. When I got out of the pool, everything seemed blurry. The whole world looked as though it had been slightly decolorized and knocked a bit off kilter. Every object I looked at now had an outline that didn't quite fit. Not to worry, I figured; splash some fresh water into my eyes, give it half an hour, have dinner, and my vision would adjust. When that didn't work, I was sure a good night's sleep would do the trick.

But when I woke up in the middle of the night and checked the big red numerals on my nightstand clock, all I could see was blurs and squiggles. Even after putting on my reading glasses, I could barely make out the time!

Next morning, panic over vision loss began to set in. I decided that if I could manage to work, that level of functionality would mean I was fundamentally okay. Unfortunately, to my great consternation, the glare of the computer screen made my eyes tear. (I know, I should have gone straight to the eye doctor, but I don't have an eye doctor, plus it was a Friday and my kids were due for the weekend, so I figured I'd fake it until Monday.) Reading actual paper copy was somewhat easier than reading from the computer screen, so I tackled a stack of research on phototherapy, which, I learned as I squinted, is the healing use of light from the Sun and from artificial sources. There was a massive printout from sunlight .orgfree.com, which posts a compendium of fascinating phototherapy research, folklore, and hogwash, all of which is best summed up by the Latin phrase, *Sol est remediorum maximum* ("The Sun is the best remedy"), from the ancient Roman writer, Pliny.

That printout had a number of references indicating that infrared radiation has a healing effect on the eyes, with several studies recommending the therapeutic benefits of looking directly into the Sun, usually at dawn or at sunset. No way was I desperate enough to look directly into the Sun for more than a second or two; I'd rather have my vision wonky than burned out. But this much seemed safe: "A natural way to expose the retina to infrared light . . . is to look in the direction of the sun, preferably at noon, with the eyelids closed. The eyelids act as filters, letting only the infrared light to reach the retina. It is necessary a long-time exposure, from 20 to 30 minutes [sic]."[1]

Look directly into the Sun at high noon for 20 straight minutes? Even with eyelids closed, wouldn't that be risky? What if I damaged my eyes forever, and by doing something that we all have been taught *not* to do? The squiggles and weird outlines were not going away, though, so I closed my eyes and tilted my face up toward the midday Sun. When I opened my eyes 20 minutes later and looked down at the pool . . . voilà! Vision restored. I exaggerate not, either in my story or in the sincerity with which I implore you never to look into the Sun with your eyes open, and not to sue me or my publisher if you do. I make no warrants, representations, or any other claims, other than the fact that gazing into the Sun, with eyes firmly closed, worked for me, and continues to help today; computer eyestrain is an occupational hazard of writers, so I redose regularly.

My God, what a worrywart I had become! Decades of public health warnings about how dangerous the Sun can be had neutered my common sense. Twenty minutes lying face-up in the Sun? Hell, I had done that hundreds of times, thousands, at the pool and the beach. Spring Break, 1973, my roommate Jack and I hitchhiked from Providence, Rhode Island, to Key West, Florida, where we found a beach, stripped down to trunks, and—for the first time after the long, cold, lonely winter—lay out in the sunshine for eight hours straight, no sunscreen. I turned crimson as the sunset; it took me half an hour to put on my shirt. For the next month I molted, shedding dead skin in 11 states. But my eyes were fine.

What other effects might the Sun have on the eyes? It is well known that excessive exposure to ultraviolet radiation, at the opposite end of the spectrum from the infrared waves that had restored my vision, can cause cataracts. I had no idea, though, that this sometimes enables those who have had their cataracts removed to see things they had never seen before. "In the first century AD, Roman doctors routinely displaced and removed irreversibly swollen and cloudy lenses from the eyes of their patients. The condition they were treating, cataract, is still with us, and still irreversible. Since 1947, it has been possible to replace the natural lens with a plastic substitute inserted into the eye. Before artificial lenses were available, however, those who had their lenses removed by surgery found that they could see into the ultraviolet; blues were clearer and richer, and ultraviolet light, energetically triggering every photoreceptor it hit, was a blueish-white wash," writes Simon Ings.[2]

Since ultraviolet levels soar at the solar climax, the Moody Sun Hypothesis would suggest that people's vision is particularly susceptible to cataracts and related effects during those peak years and far less susceptible during trough years. Moreover, this optical turbulence extends to the animal kingdom. "Insects, birds, fish and mice see shorter wavelengths than humans, well into the ultraviolet. Many flowers boast striking patterns, only visible in ultraviolet light, to attract pollinating insects," writes Ings.[3] What a far-out notion, that the roses in my neighbor's front yard might be even prettier in the UV dimension.

THE PHOTOTHERAPY DIMENSION

Closed-eye Sun-gazing reopened my eyes to the broader fact that the Sun has the power to cure as well as to harm. We used to know that, but what with all the warnings of how the Sun can harm us, we have somehow forgotten that our star is where our basic life force comes from. Consider the opening of the hallowed Hippocratic Oath (its original ancient Greek translated into English): "I swear by Apollo, the healer, . . ."

Apollo, of course, was the Greek god of the Sun. In its simplest form, phototherapy is the process of exposing the body, human or animal, to light for the purposes of healing. The light rays trigger a chain of biochemical dominoes that, when properly nudged, can have salutary consequences for the patient. "The photophysical act of light absorption initiates a sequence of actions and reactions that can lead to a remarkable diversity of physiological endpoints, for example, plant growth, animal vision, circadian rhythm and sunburn. The unique feature of phototherapy is that light acts as a powerful drug," writes Leonard I. Grossweiner in *The Science of Phototherapy*.[4]

From time immemorial, health care workers saw the Sun mostly as a source of healing, not illness. The 13th century Icelandic epic poem known as *Edda* tells us that the Vikings used to carry their sick in the springtime to the sunny mountain slopes. Grossweiner notes that "Indian medical literature dating to 1500 BC describes a treatment of non-pigmented areas of skin by applying black seeds of the plant Bavachee or Vasuchika followed by the exposure of skin to sunlight."[5]

In an article titled "Light Therapy, Circa 1939," Cristina Luigi reports that through the first half of the 20th century, phototherapy was widely employed: "Sunlit spas nestled high in the mountains became very popular among those who could afford them, and color lamps for treating a variety of illnesses were common fixtures in many rooms. Experiments on microorganisms, animals, and even humans revealed all sorts of beneficial effects of sunlight, many of which are still recognized and appreciated, such as enhanced immune function, improved musculature and even a banishment of the blues."[6]

Probably the most common current use of phototherapy is to treat jaundice in newborns. This treatment was discovered in 1857 when a nurse (whose name is lost to history) at the clinic of R. J. Cremer at Rochford Hospital in Essex, England, noticed that neonatal jaundice faded quickly in areas of the nursery that were sunlit. Neonatal jaundice severe enough to require treatment afflicts roughly 5 percent of U.S. newborns, including my godson, J. B., who looked like a little glowworm while bathed in the blue light from his ultraviolet lamps. It would take several days before J. B.'s liver cells matured enough to manufacture the enzymes that eliminate bilirubin, the yellowish waste product of the breakdown of red blood cells. In the meantime, blue light baths did the job his tiny liver was not yet up to.

Imagine a skinny wooden Tinkertoy racehorse trapped in its stable because it's too tall to duck through the door. Now rearrange its parts into a short, broad-shouldered workhorse that can walk right out, no sweat. That, in essence, is what blue light does to bilirubin molecules: it reshapes them into a form that the body can get rid of. In a process known as photoisomerization, photons, infinitesimal packets of light, penetrate the bilirubin and weaken the energy bonds that hold its molecules together. The net effect is to rearrange the molecules' atoms into a more compact configuration that allows the bilirubin to pass through the kidneys and be excreted as urine.

Now imagine that some of the Tinkertoys had been handled too roughly and then broke as a result. This would be akin to a related but different process, photoionization, which is what would have happened if the lamps shining on J. B.'s tiny body had been too intense. Overdosing newborns with blue light excites the photons to the point where they not only weaken the bonds of the atoms they penetrate, but actually knock some of the electrons clear out, ionizing the bilirubin—that is, giving it an electrical charge and therefore a whole new and potentially dangerous set of physical and chemical properties. Think jagged Tinkertoys leaving splinters in your fingers.

The modern era of controlled phototherapy started at the turn of the twentieth century, when Niels Finsen, a Danish physician, discovered that ultraviolet radiation from a carbon-arc lamp could be used to treat tuberculosis of the skin. (Arc lamps produce their light, sometimes quite intense, from arcs generated by electrical stimulation of various substances, such as carbon, in the case of Finsen's lamp. The characteristics of the light vary with the substance used. Typical fluorescent lighting used today in home and office are usually low-intensity arc lamps using mercury vapor.) It turns out that UV radiation directly damages bacterial DNA, such as might cause cutaneous tuberculosis. Moreover, when these bacterial molecules are exposed to UV light, they effectively turn against their own and become bactericidal agents. Considered the father of modern phototherapy, Finsen in 1903 received the third Nobel Prize awarded for physiology or medicine. His method of shining concentrated UV light for phototherapeutic purposes has since been adapted for use in treating acne, vitiligo (loss of skin pigmentation due to problems with the same melanocytes we were discussing earlier), and especially psoriasis.[7]

Psoriasis occurs when the immune system sends out faulty signals that accelerate the production of skin cells, resulting in crusty patches of dead skin on the elbows, knees, lower back, and elsewhere. For centuries, sufferers of psoriasis and other diseases of the skin were advised to expose themselves to midday sunlight; this worked moderately well. In the 1920s mercury-arc lamps were developed to focus UVB, the more powerful, burning range of the ultraviolet spectrum, on the affected skin areas. Basically, UVB damages and destroys psoriatic skin cells much as it damages and destroys normal skin cells.

The central principle of photochemistry is that only *absorbed* light can produce a chemical change. Light absorption depends largely on the color of the substance doing the absorbing. "All interactions of light with biological systems utilize unique light-absorbing molecular units or *chromophores* located within the illuminated tissues. In natural photobiology, the chromophores are normally present in the tissue matrix, for example, chlorophyll molecules in green plants and retinal pigments in vertebrate retina," explains Grossweiner.[8] By contrast, in the medicinal practice of phototherapy, extrinsic chromophores such as dyes and other chemical agents may be applied. For example, sometimes patients' psoriasis patches are rubbed with coal tar, which, because of its dark color, absorbs the healing light more efficiently.

The other important factor in light absorption, the opacity of the absorbing substance, gives us an unexpected glimpse into scientific genius. Most biological tissues are too cloudy and lumpy and full of odd floating bits for the rules of conventional optics to apply. Light rays entering tissues and other "turbid media" are scattered willy-nilly. So how to calculate the proper dosage of light for a given part of the body? The answer, it turns out, comes from the heavens. In the 1930s, Subrahmanyan Chandrasekhar, a cosmologist from the University of Chicago, developed a theory known as radiative transfer to calculate light propagation in another turbid medium, the universe (which, of course, is full of stars, planets, space junk, black holes, and many other inconsistencies and anomalies). The radiative transfer theory—which applies to medical phototherapy as well as to the universe—states simply that as a beam of light or other form of radiation travels, it loses energy due to absorption, gains energy due to emission, and redistributes its energy due to scattering. Writing the equations to quantify all that with precision takes rare brilliance; Chandrasekhar, having done so, won the Nobel Prize in physics in 1983. What a genius! His theory has been successfully adapted not only to tissue optics and phototherapy, our concern here, but also to applications ranging from the diffusion of neutrons in nuclear reactors, of sound waves underwater, and of radar in the atmosphere.

In the world of phototherapy, radiative transfer calculations have proved particularly useful with the introduction of a group of plant-based extrinsic chromophores called psoralens. It had long been noted that when horses, sheep, and cattle grazed upon the genus of plants known as *Psoralea* and then basked in the Sun, ulcers, infections, sunburns, skin cancers, and (in severe cases) convulsions and death resulted. Herdsmen who knew about those nasty effects perhaps wondered why God or nature had created such infernal plants. The answer, as it took centuries to discover, was to create psoralens, a family of oral and topical drugs that render the skin extremely sensitive to sunlight, thus enhancing the healing effects of phototherapy. Psoralens are far more effective chromophores than coal tar, for example, in rendering skin sensitive to sunlight, affecting not only the surface but also the (cloudy, lumpy, turbid) tissue beneath. Radiative transfer calculations are thus required to determine light dosages. These calculations, no doubt accompanied by some unfortunate clinical trial and error, led phototherapists to switch to using UVA radiation, less powerful than UVB, lest severe sunburns or skin cancers occur. (Another potent photosensitizer of this ilk is the red pigment hypericin, which is found in some insects and in plants of the genus *Hypericum,* including Saint-John's-wort, which is why users of that herb are cautioned to avoid exposure to direct sunlight.)

Despite its effectiveness, phototherapy treatment, whether through the use of lamps or through simple exposure to the Sun, has been slow to regain the prominence it attained during the first half of the 20th century. I believe this is because we are caught in a psychological double bind regarding the power of light. On the one hand, we take light too lightly, skeptical that any treatment akin to "shining a flashlight" on our body could ever really cure anything. (Unless, of course, that light is a laser beam, which is quite a different category of treatment. In phototherapy, light sources are used to stimulate comparatively subtle photochemical chain reactions, while in laser surgery tissues are forcibly vaporized, coagulated, or otherwise excised.) On the other hand is the fear, perhaps inordinate, that UV radiation is a dangerous cause of skin cancer.

Our fear of the deleterious effects of sunshine has prejudiced us against phototherapy which, when judiciously applied, clearly has enormous benefits. How odd that when it comes to seeking pleasure, we are ready to blow a wad on a Hawaiian beach vacation, but when it comes to health, the Sun is off-limits. That the Sun feels good and stimulates the body could pass as a truism in a travel brochure, but them's fightin' words when it comes to the

Western medical establishment. Consider that many Chinese hospitals today routinely move their patients outside between the hours of 8:00 and 10:00 A.M. on sunny days, so that they benefit from the healing effects of morning rays. If nothing else, the morning Sun feels good, a blessing to those who are hospitalized. In the United States, such a practice might provoke lawsuits.

THE SUNSHINE VITAMIN

In 2004, Dr. Michael F. Holick, author of *The UV Advantage*, a book that advocates careful exposure to sunlight as a way to boost health, got fired from his position as a professor in the dermatology department at Boston University's medical school. Although Holick readily acknowledges that sunburns can lead to skin cancer, he also points out that limited, regular exposure of the limbs, not the face, to sunlight safely raises levels of vitamin D, a.k.a. "the sunshine vitamin," promoting fertility and bone development and also easing depression. For that he got bounced?

Phototherapy advocates such as Holick are caught in the same sort of political double bind as those who insist that sunspots contribute to global warming: they are on the wrong side of prevailing scientific opinion regarding the Sun. At issue are not so much the facts of the matter. Few dermatologists would seriously disagree with Holick's cautious recommendations for Sun exposure. Exposing the limbs to midday sunshine for 10 to 15 minutes several times per week is enough for people of average complexion living in middle latitudes to make enough vitamin D_3 to meet their requirements. Those with darker complexions need longer exposure and may need to take supplements. What the skin doctors vehemently, viscerally object to is the fact that one of their own makes any such recommendation at all. Dr. Barbara Gilchrest, the one who fired Holick, reportedly said that she could not have "anyone who recommended sun exposure" working in her department.[9] The agreed-upon dermatological public health goal is that people should get *less* exposure to the Sun, not *more*. Consciousness should be raised regarding the *threats* from sunlight, not the *benefits*. It's the same kind of slippery slope mentality that gives Gore/IPCC conniptions when someone sensibly suggests that the changing behavior of the Sun may play some role in climate change. The bluenoses do not necessarily disagree with that hypothesis, but they do object, violently, to the resulting lack of public resolve to tackle what they see as the real cause of climate change, the accumulation of greenhouse gases.

Unlike the global warming debate, which manages to stand pretty much on its own, staying clear of extraneous issues, the phototherapy dispute over exposure to sunlight has gotten rolled into a larger and even more emotional debate about vitamin D. Vitamin D has a magical aura about it. It is unique in the panoply of human nutrients in that we can ingest it via foods or supplements and we can also photosynthesize it in our skin. As per the Moody Sun Hypothesis tenet that the Sun has significantly influenced human development, note that vitamin D evolved around 750 million years ago to help sea creatures maintain their skeletons as they moved out of calcium-rich oceans onto sunnier dry land, where calcium was scarcer and therefore had to be digested more efficiently, the primary nutritional role of vitamin D.

Funny to think that we humans are also photosynthesizers, at least when it comes to vitamin D. Because this substance can be obtained from the Sun, it is not a true vitamin, which, by definition, must come from plant or animal substances. Rather, it is a steroid hormone, of which there are five gradations, vitamin D_1 through vitamin D_5. Vitamin D_2 and vitamin D_3 are the most common and important. Vitamin D_2 (ergocalciferol) comes naturally in foods such as tuna, salmon, mackerel, and cod liver oil. It is even more plentiful in fortified foods such as orange juice and milk, a cup of which provides 25 percent of the recommended daily allowance. Vitamin D_3 (cholecalciferol) is created in high-density or "good" cholesterol when sunlight in the UVB range of the spectrum interacts with the epidermal layer of the skin. Both forms of vitamin D are considered dietetically interchangeable, although some controversy exists over whether D_2 is as broadly effective as D_3.

"Vitamin D promotes calcium absorption in the gut and helps maintain adequate serum calcium and phosphate concentrations to enable normal mineralization of bone[;] . . . it is also needed for bone growth and bone remodeling. . . . Without sufficient vitamin D, bones can become thin, brittle or misshapen. . . . [W]ith calcium, Vitamin D also protects older adults from osteoporosis," explains the "Dietary Supplement Fact Sheet" of the National Institutes of Health.[10] Another widely recognized consequence of vitamin D deficiency is difficulty digesting fats, such as afflicts those who suffer from cystic fibrosis or from inflammatory bowel disease (also known as Crohn's disease), according to the Mayo Clinic.[11]

No disagreement thus far. The trouble starts with the almost bewildering array of second-echelon, less-substantiated health claims made for vitamin D. Over the past decade, vitamin D has replaced vitamin E as the "miracle supplement," touted as effective against cancer, diabetes, arthritis, multiple sclerosis, and certain autoimmune diseases, all of which are thought to be at least partially preventable by maintaining sufficient levels of the nutrient. So prevalent have these health claims become that the United States and Canadian governments jointly commissioned the Institute of Medicine, the health arm of the National Academy of Sciences, to get to the bottom of it all.

Here's what those researchers concluded: "Despite the many claims of benefit surrounding vitamin D in particular, the evidence did not support a basis for a causal relationship between vitamin D and many of the numerous health outcomes purported to be affected by vitamin D intake. Although the current interest in vitamin D as a nutrient with broad and expanded benefits is understandable, it is not supported by the available evidence." The Institute of Medicine study examined a vast array of scientific and medical literature identified by a panel of 14 scientists and is widely considered the most objective analysis of the role of vitamin D in human health. The report of that analysis includes the following: "Scientific evidence indicates that calcium and vitamin D play key roles in bone health. The current evidence, however, does not support other benefits for vitamin D or calcium intake. More targeted research should continue. Higher levels have not been shown to confer greater benefits, and in fact, they have been linked to other health problems [such as kidney stones and tissue damage], challenging the concept that 'more is better.'" True to its findings, the institute's report lowered recommended daily consumption of vitamin D, from 1,000 to 600 IU daily, in most cases.[12]

Despite the Institute of Medicine's bottom-line conclusion that increasing vitamin D intake is counterproductive, a cadre of respected researchers beg to differ. For example, Lubna Pal of the Yale University School of Medicine declares that vitamin

D insufficiency is "pandemic." Her studies have found multiple links between vitamin D insufficiency and infertility in women, suggesting that dietary supplementation of this substance may spare many women the discomfort, trauma, and expense of conventional treatment with infertility medications.[13] Pal's research prompts some interesting speculation. If a lack of vitamin D leads to infertility in women, and a full dosage of the substance enhances women's chances of conceiving, what bearing might this have had on the population explosion of the past 150 years, coinciding as that period did with heightened solar activity? Does moving to the sunbelt make a woman in her childbearing years more fertile through increased exposure to the Sun? Perhaps there is a connection between the high birthrate of countries in the developing world and the tendency of those countries to be in sunny climes.

Similarly, male sex drive correlates closely with sunbathing and vitamin D. A major Austrian study published in late 2009 found that testosterone, which is responsible for both fertility and sex drive in men, peaks near the end of the summer in August and troughs in March, at the end of the winter; vitamin D levels follow the same ebb and flow.[14] Has the sexual revolution of the past half century or so been at least partially the consequence of heightened solar activity? Would the sexual revolution have taken place if we were still mired in the Little Ice Age? Increased geomagnetic activity, resulting in increased UVB stimulation of the skin, certainly seems to have gotten the guys going. Ditto the rising global temperatures, whatever the cause. Of course, sunspots are just one factor in the boom, along with more conventional explanations such as the shift in social mores.

A 2011 study, published after the Institute of Medicine's report, found a linkage between tuberculosis, which has a higher incidence among dark-skinned people, and vitamin D insufficiency. "Increasing vitamin D levels through supplementation may improve the immune response to infections such as tuberculosis," declares the study's lead author, Mario Fabri, who conducted the research at UCLA. Fabri adds a historical dimension to his argument: "Over the centuries, vitamin D has intrinsically been used to treat tuberculosis. Sanatoriums dedicated to tuberculosis patients were traditionally placed in sunny locations that seemed to help patients—but no one knew why this worked."[15] Similar linkages obtain for prostate cancer, according to a Johns Hopkins special report published subsequent to the Institute of Medicine study: "Sun exposure also appears to influence a man's risk of developing prostate cancer. Research suggests that men who were born in sun-drenched areas of the United States are about half as likely to develop prostate cancer later in life as men born in an area with low sun exposure. In the study, higher recreational sun exposure in adulthood also cut men's risk of fatal prostate cancer in half."[16] The Hopkins report goes on to note weaker yet still provocative correlations between living in southerly latitudes that receive year-round sunshine and lowered risks of dying from cardiovascular disease, diabetes, and colon cancer. It all boils down to the fact that people find it easier to stay fit and active, and therefore healthier, in sunny seasons and climes.

In light of the powerful health evidence in favor of regular, measured exposure to the Sun, one would think that Holick would have been applauded, not censured. We humans are counseled to consume, to pay good money and absorb extra calories and chemicals, in order to obtain our vitamin D, when what we really need is to go out, get some exercise, enjoy a little sunshine, and sure, use some sunscreen if we're going to be outside for a while. We will feel better and stay healthier longer for that time in the Sun. That's the kind of advice most of us need to follow.

"Poison is the dose," remarked Paracelcus, the legendary 16th-century Swiss physiologist. He was referring to substances such as salt, harmless (even beneficial) in small amounts, but deadly if overconsumed; water is absolutely essential to sustain life, but too much of it and it causes drowning. To understand the role that the Sun plays in our health, all we really need to do is extend the "dose" notion to include its rays. A little sunshine is necessary for body and mind to stay healthy; even more is needed by those faced with maladies such as psoriasis and/or deficiencies of vitamin D. Too much exposure, however, can burn and damage the skin, perhaps cancerously. Proper doses vary from individual to individual, and while we are a long way from personalized Sun prescriptions, it is nonetheless incumbent upon each one of us to give the matter some thought and even perhaps seek medical and/or psychological counsel. What is the right sunshine dosage for you? Me, I go nuts without sunshine and, being kind of dark-skinned, I soak it up by the kilowatt, so I'm probably not a good one to ask. And yet the experts, a.k.a. dermatologists, are so afraid of the rays that even Dr. Holick, the pro-Sun rogue, claims he never goes outside in daylight hours without sunscreen on his face. Then again, truly protective sunscreen has been on the market for only a few decades, meaning most of humanity survived prior to that point without being done in by the Sun.

10

Sunspots and Your Brain

The Sun has been screwing around with our moods without our knowing it, yanking our emotions as though we were puppets. Fluctuations in solar activity turn out to correspond intriguingly with people's outlook on life. In other words, sunspots are messing with our minds.

Anna Krivelyova and Cesare Robotti, writing for the Federal Reserve Bank of Atlanta, probe this issue in "Playing the Field: Geomagnetic Storms and the Stock Market":

A large body of psychological research has shown that geomagnetic storms [disruptions in Earth's magnetic field, usually caused by energetic impacts issuing from the Sun] have a profound effect on people's moods, and, in turn,

people's moods have been found to be related to human behavior, judgments, and decisions about risk. An important finding of this literature is that people often attribute their feelings and emotions to the wrong source, leading to incorrect judgments. Specifically, people affected by geomagnetic storms may be more inclined to sell stocks on stormy days because they incorrectly attribute their bad mood to negative economic prospects rather than bad environmental conditions.[1]

In other words, the future looks worse during a geomagnetic storm (GMS). Knowing that such a storm was taking place might well help bad-mood sufferers understand the cause of their dark feelings, thereby helping them to lighten up a little, kind of the way allergy sufferers know not to take it personally when hay fever season dampens their spirits. As economists, Krivelyova and Robotti focus primarily on how the mood-depressing effect of geomagnetic storms affects the stock market. Their conclusion, in a nutshell, is that those perturbed by such storms tend irrationally to seek out and therefore overvalue relatively riskless assets such as treasury bills.

GEOMAGNETIC DISRUPTION AND THE HUMAN MIND

Explanations for how a GMS causes such disturbances in the human mind and body tend to focus on mechanisms of magnetic interaction and its consequences. In 1992, Joseph Kirschvink, a geobiologist at the California Institute of Technology, discovered that our brains contain "little biological bar magnets made of crystals of the iron mineral magnetite.[2] Presumably, these tiny, sensitive magnets are susceptible to ultralow frequency (ULF) and extremely low frequency (ELF) electromagnetic disturbances such as might come from a GMS.

Much has been ascribed to a phenomenon known as the Schumann resonances, mathematically predicted by Winfried Otto Schumann in 1952 and experimentally confirmed and measured subsequently by numerous scientists. Schumann reasoned that the space between Earth's surface and the ionosphere (the ionized region of the atmosphere that begins about 800 kilometers above the surface would naturally act as an echo chamber for ELF electromagnetic waves. Moreover, lightning bolts would perforce disturb this echo chamber, creating highly charged resonances. Subsequent research into the resonance effects of other natural electromagnetic phenomena has come to include geomagnetic storms, which seem to have a subtle yet not insignificant impact—sufficient, a growing chorus of researchers insists, to titillate those little bar magnets in the brain.

Neil Cherry, of Lincoln University in New Zealand, contends that the scientific evidence regarding the Schumann resonance phenomenon "*strongly supports* the classification of S-GMA as a natural hazard" (italics mine). Cherry argues that the Schumann signal is highly correlated with the solar and geomagnetic activity (S-GMA) indices of sunspot number and strength of solar blasts that hit Earth. "Cherry 2002 shows that there is strong and robust scientific evidence that the human brain detects and responds to the Schumann resonance signal. . . . The reaction of the brain causes altering of the melatonin/serotonin cycle," writes Cherry.[3]

To put that into simple terms, sunspots explode with blasts that, when they hit our planet and pierce its protective magnetic field, resonate with the global echo chamber; this in turn disturbs our brains, which then fail to produce sufficient melatonin. The results of this disturbance can range from a reduced sense of well-being, to an inability to sleep, to depression or even suicide.

"The most plausible explanation for the association between geomagnetic activity and depression and suicide is that geomagnetic storms can desynchronize circadian rhythms and melatonin production," confirms Kelly Posner, a psychiatrist at Columbia University in New York City.[4] Posner explains that circadian rhythms rely on environmental cues—such as darkness—to synchronize internal clocks, and the magnetic field may be one of those cues. Disturbances in the magnetic field may desynchronize these clocks, precipitating seasonal affective disorder (SAD) and thereby increasing suicidal tendencies.

"The Sun doesn't merely drive the climate; it drives our moods and outlook. Happy people have a 'sunny disposition.' We 'see the light' of wisdom and find ignorance and evil in the shadows. . . . Our bodies slow down, our energy saps, our outlook darkens," write Steele Hill and Michael Carlowicz in their book, *The Sun*.[5] They speculate that SAD is akin to hibernation, a link between us and our winter-averse ursine cousins. Although initially dismissed as unscientific folk medicine, SAD was formally recognized in the fourth edition of the *Diagnostic and Statistical Manual of Mental Disorders*, not as a disorder per se (as its name would suggest), but rather as a recurrent seasonal manifestation of general depression due primarily to the lack of sunlight. Approximately 5 percent of the U.S. population suffers from this condition. The highest incidence of SAD is found in regions with long, dark winters; almost 1 in 10 individuals living in areas such as Scandinavia and New England are said to suffer from it. Typical symptoms include depression, undue pessimism, suicidal thoughts, and excessive sleep, all occurring primarily or exclusively on a seasonal basis. Treatments include bright-light therapy, antidepressants, talk therapy, and melatonin hormone supplements.

Although normally thought of as a winter phenomenon, there is also a much rarer summer version of SAD—a version in which not a *deficit* but a *surfeit* of sun causes problems—afflicting less than 1 percent of the general population. Not surprisingly, SAD/summer is more prevalent in sunnier climes. For example, about 1 in 55 Floridians, almost 2 percent, experience SAD/summer, symptoms of which include anxiety, insomnia, weight loss, and increased sex drive. Researchers are still puzzled as to the trigger; melatonin levels are thought to *not* be involved with this variant of SAD. Acute sensitivity to heat and light is, no surprise, a prominent factor. Warm nights apparently disrupt some sufferers' sleep cycle, and bright sunshine gives others the sense of being imprisoned in glare. Treatments include avoidance of the outdoors whenever possible, cooling regimens, and antidepressants, some of which lower patients' body temperature at night. Because SAD/summer is so rare and esoteric, few studies of its broader consequences have been undertaken. Nonetheless, folks who suffer from this disorder would be well advised to be especially careful about making any major life decisions during the summer.

THE IMPACT OF ELECTROMAGNETIC FLUCTUATIONS

One can readily intuit how long, cold periods of darkness might dim the spirit, and one can even empathize with those overwhelmed by the summer Sun. But random fluctuations in Earth's magnetic field? If you took a magnet and rubbed it all over your head, would it scramble your thoughts? Common sense says that the brain, with its delicate electrochemistry, is bound to be vulnerable to external electromagnetic interference such as might come from extreme solar outbursts. It's the same homespun logic which argues that the Moon, whose gravitational pull governs the tides, must also have some effect on the human body and brain, which are made up mostly of water. Radiation from the Sun has a significant electromagnetic component, as do the impulses transmitted by our nervous system and the thoughts that course through our minds. It only stands to reason that solar irregularities would therefore rattle our noggins.

But how exceptional does a solar outburst have to be to penetrate the skull? Are thinner skulls, such as those encasing the brains of young children, more vulnerable? How to factor in the state of Earth's protective magnetic field at the time of the Sun–brain encounter? As has been noted, from time immemorial sunspots have waxed and waned significantly over an 11-year cycle, solar activity varying in frequency and intensity by a factor of three times or more between trough and peak. Presumably, our brains are used to this boom-and-bust pattern of solar behavior. Wouldn't it take a radical departure from this norm, one way or the other, to truly affect how we think?

"There seems little doubt that the brain responds to electromagnetic fields—coils that generate electromagnetic fields can trigger muscular twitches when placed over a human skull," observes the *New Scientist* study.[6] Indeed, since 1985, a process known as transcranial magnetic stimulation (TMS) has been used to help remedy brain disorders such as schizophrenia, migraines, and major depression. The TMS device sends an electromagnetic pulse into the skull, essentially resynchronizing neuronal firing.

The fact that targeted electromagnetic skull "injections" can affect the brain does not necessarily mean that the brain is also susceptible to the much subtler fluctuations caused by geomagnetic disturbances. Nonetheless, the basic connection is clear: every time a neuron fires in the brain, it disturbs the electromagnetic field surrounding that organ. Conversely, disturbances in the surrounding electromagnetic field are thought to affect neuronal firing. (This is one of the reasons that people worry about what excessive cell phone usage does to the brain.) In an article that queries, "Could certain frequencies of electromagnetic waves or radiation interfere with brain function?" Amir Raz, a clinical neuropsychologist from Columbia University, answers his own question: "Definitely. Radiation is energy and research findings provide at least some information concerning how specific types may influence biological tissue, including brain tissue. . . . For example, a recent report suggests that the low-intensity electromagnetic fields of geomagnetic storms—disturbances in the earth's magnetic field caused by gusts of solar wind—may have a subtle but measurable influence on the suicide rates in women."[7]

In a different article looking at disruptions in Earth's magnetic field and suicide, New Scientist magazine surveyed various studies and indeed found numerous significant correlations, if not definitive causal relationships, between geomagnetic storm spikes and increases in suicide rates (and therefore, presumably, in depression).[8] From the northern Russian city of Kirovsk to Australia and South Africa, researchers noted the GMS–suicide connection.

However, extra caution must always be taken when working

with suicide statistics. In Catholic countries, suicides are typically underreported because the Church, which considers suicide a sin, denies the victim a proper religious funeral mass and burial. This stance sometimes prompts families of suicide victims to misrepresent the circumstances of a loved one's demise. By contrast, suicide statistics are meticulously maintained and reported in Scandinavian countries, where people believe that society has a right to such information.

Supplementing suicide data are numerous other studies linking geomagnetic disturbances to the incidence of clinical depression, admissions to psychiatric hospitals, and even disturbances in fetal heart rates. Geomagnetic storms may rage without catching the attention of the average person, but apparently they have an impact nonetheless.

According to the previously cited Atlanta Federal Reserve study, only 10 to 15 percent of the population is considered susceptible to geomagnetic outbursts. Presumably those most vulnerable to SAD are heavily represented in this group. That would likely skew GMS/SAD incidence toward regions with long, dark winters, the higher latitudes of both hemispheres. (It would be interesting to see a study relating GMS/SAD vulnerability and skull thickness.) Ultimately, the question seems to be one of degree: How big a threat is sunspot confusion? Is the malady truly a driver of suicide rates, or does it simply cause a periodic melancholy tinge? Maybe sunspot confusion is like peanut allergy, statistically rare but so dangerous to those vulnerable few that broader precautions are warranted.

The Sun's moods permeate our own, whether we are among the 10 to 15 percent of the population particularly susceptible, or just part of the hapless majority who have to deal with the sunspot confusion of others. The feelings of insecurity engendered by this phenomenon almost certainly extend to aspects of human endeavor other than finance and might well affect the whole range of decision making, from the societal level on down to the personal. But what to do with this knowledge about GMS/SAD? Simply being aware of the possibility that one's moods might be negatively affected, through no fault of one's own, is probably the best way to mitigate this condition. Same goes for understanding that others might be thus afflicted. Taking melatonin supplements might not be a bad idea if one finds one's sleep unusually disturbed during periods of high geomagnetic activity. Those scientifically inclined might endeavor to develop a reliable means to test an individual's relative susceptibility to GMS. Wall Street firms, defense departments, and other institutions whose proper functioning depends critically upon objective, "mood-free" risk assessments might want to screen their applicants for this susceptibility.

Concomitantly, risky assets such as start-up companies and their high-tech innovators tend to be undervalued during heavy GMS periods. However, even the blue chips may take a beating. As the Atlanta Federal Reserve study noted, "Unusually high levels of geomagnetic activity have a negative, statistically and economically significant effect on the following week's stock returns for all US stock market indices." Conversely, the researchers found "evidence of substantially higher returns around the world during periods of quiet geomagnetic activity."[9] It should be noted that the stock market did indeed decline as predicted during the week following the highly publicized geomagnetic storm that hit on January 24–25, 2012. According to the Associated Press, stock market indices had done well for most of that month. However, both the Dow Jones Industrial Average and the Standard and Poors 500 indices reversed direction, flattening out or outright declining during the week beginning January 25, 2012. "An unexpected drop in consumer confidence dragged stocks down on the final day of the month," according to the Associated Press.[10] Were investors' nerves jangled by geomagnetism?

How many important decisions over the course of history have been made under the influence of solar interference? Has the Moody Sun Hypothesis been shaping history in mysterious ways? While it is as yet impossible to determine what percentage of the population might be strongly influenced by spikes and/or troughs in sunspot activity, even if one assumes the proportion to be small, that does not necessarily mean that the overall impact would be negligible. Sometimes it takes just one or two exceptional outcomes to affect everyone. For example, if the president or other major political leaders were vulnerable to GMS/SAD influence, key decisions could be negatively influenced. Gawker blogger Adrian Chen makes the argument that President Obama suffers from seasonal affective disorder and that it puts him in a sad, lonely mood.[11] If that's in fact the case, knock wood, for his sake—and ours—that this disorder, which usually sets in around the beginning of November, is not exacerbated by the 2012 election results from that month.

I think of GMS/SAD as a trim-tab factor, a trim tab being part of the rudder of a ship—a rudder on the rudder, if you will. Every 11 years or so, solar outbursts cause the great ship of state's trim tab to turn slightly toward pessimism and despair. Of course, no trim tab can override a roaring current; the powerful optimism generated by peace and prosperity will not be undone by a few solar hiccups. In this way, GMS/SAD is destined to go against the flow. In our culture, at least, optimism is doctrinaire. The whole economy seems predicated on rising consumer confidence and rosy expectations. But perhaps, in the grand scheme of things, the occasional touch of solar pessimism is helpful as a brake against inflated expectations.

Sunspot activity may well prove to be influencing our thoughts and our moods in ways we never expected. Psychologists might consider studying a sample group's decision-making performance during a solar climax compared with its performance during the ensuing solar trough. (Factoring out all the other variables affecting judgment and cognition would be challenging, therefore time-consuming, therefore grant-deserving.) One line of inquiry might be to monitor the activity of brain waves via positron emission tomography (PET) scans and other such means during geomagnetic storms. The regions of the brain that show the most activity could be expected to be at their peak every 11 years. As we have seen, spikes in solar output effectively block cosmic rays from reaching Earth, and troughs in solar output allow these rays—energy input *not* from the Sun—to flood in. Some brains may be particularly sensitive to cosmic fluctuations; others, especially immune.

GEOMAGNETIC DISRUPTION AND INTUITIVE PERCEPTION

Half a century of research in Russia suggests a strong correlation between reduced cosmic-ray exposure and enhanced psychic/intuitive perception. If the Russians are right, every time Usoskin's graph spikes—more solar activity and thus fewer cosmic rays—so does our collective ESP. To investigate this matter, in 2006 I visited Akademgorodok, an extremely advanced scientific research city in Novosibirsk, Siberia, where I learned that almost half a century of Soviet/Russian academic inquiry had been devoted to understanding the relationship between solar output, cosmic rays, and psychic/intuitive perception. Alexander V. Trofimov, general director of the International Scientific Research Institute of Cosmic Anthropoecology (ISRICA), and head of the laboratory at Helioclimatopathology Scientific Center of Clinical and Experimental Medicine of the Siberian Branch of the Russian Academy of Sciences, has firmly established that cosmic rays can interfere with psychic/intuitive perception. Just as cosmic rays cause clouds to form in the atmosphere, so they also serve to cloud certain nonlinear aspects of judgment, according to Trofimov.[12] Conversely, reduced cosmic-ray activity, such as results from heightened solar activity, has been significantly correlated with enhanced psychic/intuitive perception. I have since revisited Akademgorodok to inspect several devices the Siberian scientists have developed to block out cosmic-ray interference and therefore enhance certain forms of perception and communication. As I reported in *Apocalypse 2012*, preliminary experiments conducted using these shields to enable telepathic communication have yielded promising results.

Another fascinating area of the Siberian research program is the effect of cosmic rays, and the blocking of such rays by heightened solar activity, on water. The scientists I met with firmly believe that water's physical structure varies with levels of exposure to cosmic rays, and that the ingestion of high- or low-exposure water has commensurate effects on the human body, particularly the brain (which is 70 percent H_2O, compared to 60 percent of the average adult body). Babies' brains have an even higher water

content and therefore should be considered particularly susceptible to cosmic-ray fluctuations.[13]

Comparing findings from the Atlanta Federal Reserve and
the research institutes of Akademgorodok raises some interesting questions. Geomagnetic effects issuing from spikes in solar
activity have been linked to a degradation of the kind of decision making necessary to succeed in the stock market; at the
same time, such spikes have been linked to the enhancement
of psychic/intuitive thinking. Is this a right-brain/left-brain
phenomenon? Are some folks more susceptible to geomagnetic
disruption, while others are more sensitive to cosmic-ray interference? How effective are the Russian devices at shielding users
from cosmic rays and at simulating the effects of solar spikes, and
of what practical use in stock market investments and other areas
of human activity might these devices be? How many profoundly
important decisions over the course of recent history have been
(unwittingly) made under the influence of solar peaks or valleys?

Perhaps our search for Sun–mind connections will ultimately
disappoint, just as happened in the decades following the Carrington event of 1859, that spurred overblown expectations
that everything from agriculture to the stock market could be
explained by solar activity. But this time, our technology is extremely advanced, exquisitely sensitive—precise enough to register the subtlest brain waves, yet bold enough literally to pierce and
analyze the substance of the Sun. New Sun–mind connections
are sure to be found. We may even come one day to reunderstand
ourselves as "Sun-heads," beings who are mentally "plugged in"
to the Sun—wireless receptors, as it were, of our star's moods and
special knowledge.

FUTURE

Behind the glass door of the bookcase in the apartment where I grew up in Park Slope, Brooklyn, was a volume entitled *The Greatest Story Ever Told*. Was the book really so terrific? I read the first few pages, then announced to my mother that the author was just bragging about how good the book was. She explained that the title referred less to the quality of that book per se than to the story it retold—the life, death, and resurrection of Jesus, certainly one of the most popular and powerful sagas of all time, the sacred story I was raised on. Since those boyhood days in the 1960s, a competing greatest story ever told has emerged: the Big Bang birth of the universe 13.7 billion years ago and the subsequent creation of our world. Did our world begin with "the word," as John teaches in his Gospel, or did things start with the discontinuity—the infinitely small, infinitely explosive wrinkle in whatever it was that existed before space-time—as Big Bang cosmologists now contend? The essential difference is that discontinuities arise on their own, but words cannot utter themselves.

How it all ends is, of course, an equally spirited debate, with hypotheses ranging from a protracted battle between the fire-and-brimstone God and Satan à la the New Testament book of Revelation, to T. S. Eliot's "not a bang but a whimper," the poetic analog of the cosmologists' conclusion that the universe will flicker out slowly, a victim of its own entropy. No one, to the best of my knowledge, has seriously suggested the Mother of All Blackouts as an end-time apocalyptic scenario, so far be it from me to toss another black hat into the ring. But this much I will tell you: running out of electricity for months or years may not be the literal end of the world, but all hell is going to break out soon if we don't protect our power supply from solar blasts.

11

The Sun Will Soon Short Out the Electrical Power Grid

The countdown to the end of civilization as we know it began on August 28, 1859, when Earth was engulfed by the first of two mammoth plasma fireballs. Bloodred auroras flashed around our planet from both poles to the equator on and off for 42 hours. Then, on September 2, 1859, an even bigger fireball hit. Rogue electrical currents surged from the atmosphere through much of the world's 200,000 kilometers (124,000 miles) of telegraph lines, disrupting service in all four hemispheres; some operators were shocked unconscious. Campers in the Rockies read their books at midnight. Magnetometers in Russia blew a fuse. Sparks flew, fires started, compasses spun, ships were thrown off course—all by what is now known as the Carrington event, named after Richard Carrington, the British astronomer who telescopically observed and then sketched the enormous sunspot that belched out the fireballs.

The Moody Sun Hypothesis can trace its intellectual roots back to this moment, since it was the first time that the connection between the behavior of spots on the Sun and turbulence on Earth was established scientifically. The initial wave of research following the Carrington event focused heavily on relating the 11-year solar cycle to cycles in nature and economics. For example, if a particular river flooded every decade or so, scientists would hypothesize a sunspot-cycle connection, then scour the water-level logs for supporting data. Or if grain prices crested or fell with some regularity, corresponding peaks and troughs in sunspot activity were searched for. This approach yielded lots of correlations and coincidences—plenty of smoke but no smoking gun. (Correlations are the statistician's equivalent of circumstantial evidence; rarely can they be used to prove something beyond a reasonable doubt, but when enough of them pile up, skeptics just have to cry uncle. Correlations indicate a *relationship* between two variables but do not necessarily signify *cause and effect*. Say, for example, that a high percentage of felons were found to have eaten regularly at fast-food restaurants. This does not mean that eating at fast-food restaurants in any way *causes* individuals to commit felonies.)

By the turn of the 20th century, the pile of circumstantial correlations between sunspots and various phenomena had grown impressively thick. For example, a study of Bordeaux wines from 1845 to 1915 concluded that the best vintages neatly matched the peak years of the solar cycles.[1] The most mediocre vintages occurred at the trough of the cycle. "Waves" from the sunspots somehow stimulated the health of the grape crops beneficially; that was the theory, though it still remains unproven. The Moody Sun Hypothesis would now be entering its second century if anyone had been able to establish a solid causal link between the Sun's behavior and the quality of this magical and, to some, sacred fluid. Roland Barthes refers to wine as the "sap of the sun and the earth," adding that "it is above all a converting substance, capable of reversing situations and states, and of extracting from objects their opposites—for instance, making a weak man strong or a silent one talkative. Hence its old alchemical heredity, its philosophical power to transmute and create *ex nihilo*."[2] The hypothesis might also have received a boost had my antecedents been able to sustain their contention that a *lack* of sunspots is what sank the *Titanic*. A two-decade dip in solar activity around the turn of the century coincided with a dip in sea surface temperatures, a frigid condition accompanied by the movement of icebergs southward from Greenland into shipping lanes that traditionally had been ice-free. Thus the *Titanic* tragedy, or so some sunspot scholars of that era argued.

Sunspots came to be offered up as the cause of pretty much everything that anyone felt needed to be explained: "Over the next two decades, dozens of papers appeared relating changes in the sun to variations in the earth's temperature, rainfall and droughts, river flow, cyclones, insect populations, shipwrecks, economic activity, wheat prices, wine vintages, and many other topics," write Douglas V. Hoyt and Kenneth Schatten.[3] Hypotheses were aplenty, but proofs remained elusive. So, like Vinland, the ill-fated Viking colony in North America, this first-wave attempt at understanding how changes in the Sun's behavior shaped life here on Earth eventually died on the vine.

Sunspot scholarship was done in almost single-handedly by William Stanley Jevons, a brilliant but discredited economist. At the turn of the 20th century, Jevons failed spectacularly to demonstrate a connection between sunspots and the economy. His reasoning was sound enough on its face: solar activity affects crops—specifically, corn—which in turn affect the rest of the economy. More abundant (and therefore cheaper) corn is good for the economy because it leads to lower food prices in general, resulting in more income left over to spend on other things. (Scary to note that current world food prices have been rising steadily with the growing demand for corn-based ethanol.) Jevons's book *Investigations in Currency and Finance* contained three essays relating sunspots to commodity prices, with the goal of explaining boom-and-bust economic cycles as a function of solar activity.[4] The problem with his analysis was that solar cycles run about 11.11 years, and economic cycles run, according to Jevons, about 10.4 years. What might seem a minor discrepancy quickly compounds into major error, so Jevons attempted vainly to recalculate, a.k.a. fudge, the sunspot cycle as being a year shorter than, by all scientific accounts, it is.

Jevons's methodology was widely denounced—so vehemently, in fact, that few academics have since been brave or masochistic enough to carry on the work of this pariah. A pity, because those who sensed that sunspots are potentially important to everyday life were right beyond their wildest nightmares. They were just ahead of their time, and a little off base. Forget corn crops and cruise ships. Those researchers should have paid closer attention to the actual physical havoc wrought by the Carrington event on the telegraph system, now sometimes called the Victorian Internet. Although modest at the time, that damage was a tip-off of infinitely graver threats to come, such as the one to the electrical power grid that suckles our civilization.

SOLAR EMP

Those who crave a soupçon of danger in their relationships may be pleased to learn that the Sun will probably attempt to kill us all in the near future—that is, before the normal lifespan of some or most of us alive today would otherwise play out. As noted in the introduction, solar blasts of plasma radiation such as the Carrington event are going to knock out our electrical power grid, leaving up to 100 million Americans and countless others around the world without electricity for months or years. This is according to *Severe Space Weather Events: Understanding Societal and Economic Impacts*, the National Academy of Sciences report I cited in the introduction. The Academy's 2008 report, prepared in collaboration with NASA, finds that solar blasts the size of the ones that hit Earth in 1859, 1909, or 1921 (before electrical power grids existed) would today cause the Mother of All Blackouts, and the results would be cataclysmic. (Those little plasma burps were nothing compared to the ones that melted the most recent Ice Age. The next time we get hit with one of *those*, civilization will go up in smoke and steam.) As examined in my recent book *Aftermath: A Guide to Preparing for and Surviving Apocalypse 2012*, little or no electricity means little or no telecom, no municipally provided water or fuel (given that the pumps are electric), no refrigeration for fresh foods or medicine, and no banking, law enforcement, or military security—indefinitely. Major metroplexes from New York to Chicago to Seattle will likely become uninhabitable the next time we are hit by a Carrington-size plasma fireball, according to the above-cited report issued by the leading scientific body in the world. Mad Max?

Freeze, if you haven't already done so. Go numb with fear. Denial is a very effective form of defense, but one can hold one's breath for only so long. Now exhale and convince yourself that this is one of those worst-case scenarios that will always remain just that, a case cooked up to scare the bejeezus out of you. Everything is so exaggerated these days, with expert after expert flapping their gums about some weird thing that's going to kill you. Besides, didn't I write earlier in this section that the last wave of sunspot scholarship yielded basically nothing? No sense in getting one's knickers in a twist over a theory that crapped out a century ago.

Or perhaps you would prefer to downplay the whole blessed thing. Blackouts aren't so bad—kinda cozy, in fact, candlelight and all. Might be just what we need to slow down, step back, and reassess the mad, high-tech dash of 21st-century existence. There will be inconveniences, sure—like when the city sewer system backs up because all the pumping stations lose electricity and go offline. That will be messy, no doubt, especially when the toilets start backing up in bathrooms at home. Okay, so it won't all be a bed of roses, but we are hardy folk. We can live off the land, by our wits, make do, survive. But it does seem as though we are begging for annihilation, electrical power grids crisscrossing our continents like so many lightning rods. Lucky for us that Sol is not more like Thor, who could not resist tossing a few thunderbolts in our direction. Sol, as we understand our sustaining star, knows nothing of either temptation or restraint. God save us all.

We should pray for the Carrington event never to happen again. When the Sun spits too much plasma—highly charged, magnetically capricious energy in gaseous form—in our direction, it can scramble the power grid, even short it out entirely. Here's how the assault would go down. Sunspots explode and shoot fireballs through space, some of which strike Earth. If the incoming fireballs are really enormous, they can break through our planet's protective magnetic field. This mighty impact creates pulses of what is known as solar EMP (electromagnetic pulse)— in other words, high-intensity bursts of electromagnetic radiation. These EMP bursts spark geomagnetic storms in our planet's upper atmosphere. Geomagnetically induced currents (GICs) from these massive atmospheric storms then infiltrate and sometimes disrupt the complex North American bulk power system and other high-voltage power systems around the world.

And this isn't just doomsday scaremongering: "For many years it has been known that [geomagnetic] storms have the potential to cause operational threats to bulk power systems; both contemporary experience and analytical work support these general conclusions. . . . [T]hese assessments indicate that severe geomagnetic storms have the potential to cause long-duration outages to widespread areas of the North American grid," according to a 2010 report jointly sponsored by the North American Electric Reliability Corporation (NERC) and the Department of Energy.[5] The good news is that NERC, a private corporation that nonetheless sets the reliability standards for the nation's power grid as a whole, is aware of the problem. The very bad news is that NERC in fact has been aware of this since 1990, when its board of trustees published a high-sounding resolution confirming the danger.

But except for the occasional reference, as in the quotation above, NERC has done virtually nothing to address the problem it so compellingly identified. Those corporate bureaucrats squandered 20 years of opportunities to defend the way of life we hold so dear.

HEIGHTENED EMP VULNERABILITY

Today's intricately interconnected, hyper-technophilic global civilization is more vulnerable to solar EMP assaults than the world was in 1990, precisely because of all the technological progress we have made since then, virtually all of which depends on a stable electricity supply. Think, for example, of how lost many of us would be without our GPS systems, which rely on a navigational network that would certainly be disabled if the grid went down. Back in 1990, comparatively few of us had GPS capability and instead kept maps in the glove box. Even then, though, many cars had electronic ignitions that would have been disabled by solar EMP. Regress another 20 years to 1970, before the widespread adoption of electronic ignitions, and most cars would have been unaffected by a Carrington-type event. Make your way all the way back to 1859—the date of the Carrington event—and there was little the solar blast could do to harm what was essentially a pre-electric society. The closest thing to a power grid was the telegraph system, which, as noted, got scrambled up pretty good.

Consider these cautionary words from the 2008 *Critical National Infrastructures Report* prepared by the Electromagnetic Pulse Commission, a bipartisan entity headed by Dr. William Graham, who discovered the EMP phenomenon in 1962:

The physical and social fabric of the United States is sustained by a system of systems; a complex and dynamic network of interlocking and interdependent infrastructures . . . whose harmonious functioning enables the myriad actions, transactions, and information flow that undergird the orderly conduct of civil society in this country. The vulnerability of these infrastructures to threats—deliberate, accidental, and acts of nature—is the focus of greatly heightened concern in the current era, a process accelerated by the events of 9/11 and recent hurricanes, including Katrina and Rita.[6]

During what proved to be the last opportunity for the United States to conduct atmospheric testing of nuclear weapons, the EMP from a 1.4 megaton bomb detonated at an altitude of about 250 miles over the mid–Pacific Ocean knocked out streetlights, fuses, and phone service in parts of the Hawaiian Islands 800 miles to the east. A bomb of similar magnitude detonated about 250 miles above Kansas—in other words, over the heart of the country—would destroy most electronics in the continental United States. The same can be said for a repeat of the Carrington event.

No one knows when the next Carrington-scale impact will occur. However, as noted, we do know that the next solar climax—the next "red zone" when Sun storms will be at their most frequent and intense—is scheduled to commence at the end of 2012. Perhaps coincidentally, that timeframe is strongly associated with the ancient Maya, who attached special importance to the behavior of the Sun, particularly when it eclipsed—that is to

say, interposed itself between Earth and the dead center of the
Milky Way galaxy. Such eclipses occur once every 5,125 years,
with the next one coming on 12/21/12, the winter solstice. The
Maya attributed great significance to such cycles, although there
is no direct evidence they believed that the end of this or any
other cycle would also mean the end of human existence. Thank-
fully, there is even less evidence that such a prediction would
prove correct. However, the Maya did seem accurately to foresee
an apocalypse, in its original Greek meaning of "lifting the veil"
to see deeper truths, perhaps terrifying ones. The veil now being
lifted is the one that has obscured the life-and-death importance
of the Sun itself. Solar cycles have spiked every 11 years or so
from time immemorial, but this is the first climax we will have
experienced armed with the understanding that Carrington-scale
blasts from the Sun have the power to knock out the electrical
power grid that supports our civilization. A terrible truth indeed,
and one that, perhaps coincidentally, the ancient Maya have
helped us discover.

12

A Simple Way to Protect Our Future

One of the freakiest things about the Moody Sun Hypothesis is the incidental insight it sometimes provides into human character. As we saw in the previous chapter, there is a pretty good case that fluctuations in the Sun's behavior, à la the Carrington event, could determine—sink—our future. A damn good case. I have spoken on the subject of the Mother of All Blackouts dozens of times in front of all sorts of audiences, and generally people respond intelligently and with genuine concern. Oftentimes folks will attempt to compare and contrast the threat to the power grid to other catastrophes, such as those that result from climate change and/or military attack. Others immediately relate it to blackouts they have experienced, sometimes with fond memories of cozying together for the duration. A smaller percentage can normally be expected to pooh-pooh the whole idea, either by rating the odds of such an occurrence as vanishingly small or by opining that the powers that be must be handling the problem and have not informed us of that for reasons of national security. All of these responses seem reasonable to me, or at least within the realm of possibility. Where people really go off the rails, though, is when they are informed that there is a solution—actually, a pretty simple and cost-effective one that would prevent Mama Blackout from darkening our future—and fail to act on it.

To understand that solution, we have to have a little background. Transformers are the weak links of the electrical power grid and therefore of the civilization that depends on that grid, so we must protect these devices. Protecting transformers means protecting their ability to receive a large volume of current—for example, 750 kilovolts of AC (alternating current)—and step it down ("transform" it) into smaller, more user-friendly amounts— say, five lines of 150 kilovolts each. Smaller transformers then step the current down even further along the line until end users, such as homes and commercial buildings, receive their current in lines of 120 to 240 volts each. Divvying up current is never a perfect process, so there's always some energy lost to heat, which is why transformers tend to run hot (though not nearly so hot as if blasts from the Sun fried them). At only a few volts per square meter, the geomagnetically induced currents from solar EMP pose no direct threat to our bodily health. Most of us wouldn't feel a thing. However, these mini-shocks of DC (direct current) are more than enough to short out the largest transformers, which operate on AC protocols. (In alternating current systems, the direction of the electron flow changes periodically. In direct current systems, the flow is in one direction only. The sudden injection of direct current can disrupt the alternation rhythm, perturbing the AC system.)

Unfortunately, it's not just a matter of sending out repair crews or even the Army Corps of Engineers to restore power after EMP damage. The largest and most vulnerable transformers, which weigh 100 tons or more, could not be fixed after such an event because electromagnetic pulses would fuse their copper windings solid. The transformers would have to be completely replaced. Currently, there is a three-year waiting list on the world market for the heavy-duty transformers required to handle the mega-voltages whizzing around our superpowered grid. Compounding the problem is that these devices must be customized to the exact specifications of their spot on the grid, so it will not do simply to mass-produce and stockpile them in advance. Then there is the fact that large-capacity transformers are no longer manufactured in North America, putting us at the mercy of the production schedules and political agendas of China, India, and Brazil.

If not replacement, prevention. Just as we routinely protect our computers and flat-screens with surge suppressors, so must we use those devices to protect the power grid. Basically, the idea is to place a surge suppressor about the size of a washing machine between the ground and each major transformer. Any GICs caused by solar blasts would be blocked by the surge suppressor before any harm was done to the transformer, and therefore to the grid. There are approximately 350 transformers throughout the United States and Canada that would need to be shielded for the grid to remain basically operational. The cost estimate for this retrofit ranges from $1 billion to $1.5 billion. A lot of money, to be sure, but still less, for example, than 1 percent of the AIG bailout money!

"It's just the opposite of how we usually think of natural disasters. Usually the less developed regions of the world are the most vulnerable, not the highly sophisticated technological regions," says John Kappenman, an electrical engineer from Duluth, Minnesota, who, as profiled in my recent book *Aftermath,* has been leading the surge suppressor crusade.[1] Not that the Sun has any religious bias or intentions, but the paradoxical nature of the solar threat, in which the poorest, least technologically advanced areas of the world would probably be affected far less dramatically than the richest, does remind one of the biblical prophecy, "In the same way, the last will be first, and the first will be last. For many are called, but few are chosen" (Matthew 20:16, International Standard Version). And if there is anyone in this world who has been both called and chosen, it is Kappenman, who for the past 30 years has been campaigning to protect the power grid from solar blasts.

One would think that people would be happy to know that there is a way to avoid the cataclysm of power grid collapse, but in my experience that's just not the way things work. For some, the existence of the solution seems to detract from the romance of the threat, as though we were conditioned to respond only to the most extreme dangers. Does the fact that a problem can be solved somehow make it less important? Have we so much faith in our system that we blithely assume everything that *can* be solved *will* be solved as a matter of course? Other people seem inconvenienced by the whole notion of surge suppressors, perhaps expecting to be hit up for a donation or a tax surcharge (on top of which they might be required to sit through a lecture full of technical details). Then there are those—fortunately—for whom the existence of a solution makes the danger real, moving it into bright consciousness, out from the back-room mental compartment labeled "Denial" reserved for all those doomsday things that are too horrible to think about and are therefore never really going happen (such as all-out nuclear war, the events depicted in the biblical book of Revelation, that sort of thing)—or if they do happen, we will all be in the same (sinking) boat, anyway. The fact that surge suppressors could in fact head off "Gridmaggedon" immediately transfers the whole idea from "Denial" to "Head-ache," "Flashing Red Light," "Duty," or whatever other name one might give the compartment that includes other pressing issues such as "Get the leaky roof fixed before it falls in on your head."

Still, the greatest challenge in communicating the fact that there is something we can do to save the power grid infrastruc-ture (and therefore the way of life we hold so dear) is getting people to pay attention in the first place. If human psychology were such that repetition was the building block of emphasis— that saying something *five* times was reliably more emphatic than saying it *four* times, which in turn was more emphatic than *three*—I might fulfill my contractual obligation to provide readers of this book with meaningful content of 70,000 words in length simply by repeating the phrase "surge suppressors" 35,000 times. As we all know, however, a lot more than repetition is required to make a point sink in. It takes a story, a compelling narrative. But why should that be? Why do we need stories when we already have facts? Shouldn't it be enough that civilization's infrastruc-ture is mortally threatened, that daily life could be in shambles if we do not defend ourselves against Sol's blasts? We have an inordinate, one might even say sick, need for plot, character, and action, for an engaging and plausible ordering of events. Is our addiction to stories a flaw in our cognitive development? Perhaps it is an evolutionary holdover from oral tradition, a form that, rendered largely without supporting documentation, was heavily beholden to the dramatic skills of the teller. Whatever that need's origins, we seem to have developed an abnormally large sweet tooth for story magic, to the point where if we are not rewarded with a dollop of narrative dessert for every bite of factual protein, the meal is not satisfying.

So here's a story that will blow your fuse. In June 2010, the United States House of Representatives voted to authorize the Federal Energy Regulatory Commission, the governing body for the nation's utilities, to proceed with the process of installing the above-described surge suppressors. The bill, known as the GRID Act (short for Grid Reliability and Infrastructure Defense Act), would have allowed the utility industry to pass along costs to con-sumers, who would probably have had to pay $5 to $10 extra per household per year. The House vote in favor was *unanimous*— this at a time when most Washington politicians wouldn't cross the aisle to perform the Heimlich maneuver! But rather than following the House's affirmative lead, the Senate took its best shot at short-circuiting civilization and subverting our way of life by stripping out all language relevant to protecting the power grid from solar storms; even this watered-down version was scuttled, and never came to the floor for a vote. This change was thanks largely to the efforts of the members of the Senate's energy sub-committee, who apparently wanted to protect the utility industry from any inconvenience.

(The utility sector is not noted for innovation, which the House measure would have required. In 2009, that sector received a grade of D+ from the American Council of Engineers, which censured a lack of investment in new modes of power generation and transmission. The electrical utility industry currently spends a lower proportion of its revenues on research and development than just about any other industrial sector, including pet foods.)

In August 2010, I wrote a guest editorial op-ed piece for the *New York Times* imploring the Senate to break the impasse and move forward with the House plan to protect the power grid.[2] No such luck. At this writing, no discernible progress has been made—a real pity, since, by scientific consensus, the next solar climax, the time when solar storms will be at their most frequent and ferocious, is due to begin in late 2012 and extend through much of 2013. Note well that this does *not* necessarily mean that we will be hit by a Carrington-scale solar blast during the upcoming climax, simply that the next "red zone" is imminent. Divining when the next "big one" will hit is a pseudo-statistical exercise. Experts in the field tend to refer to the 1859 Carrington blast as a once-in-a-century event, while blithely ignoring the fact that blasts of comparable magnitude hit in 1921 and, recent evidence suggests, in 1909. Three data points do not a rigorous survey make. Suffice it to say that such a catastrophe could descend upon us at any time, particularly at the peak of any 11-year solar cycle.

In September 2010, I attended an 18-nation conference at Parliament in London—the Electric Infrastructure Security Summit—with special emphasis on the threat posed by solar EMP to the world's electrical power grids. Although this meeting and a special full-day closed-door workshop the day after were frightening and revelatory, attendees had to pledge never to reveal what transpired there. Beyond disclosing that representatives from major societal sectors such as energy, transportation, military, and health care had gathered to coordinate emergency plans (each representative undoubtedly experiencing a nervous shudder at the thought that a debilitating solar blast might hit during the 2012 London Summer Olympics, or "Apocalympics 2012"), the most constructive, best summary of those proceedings that I am at liberty to provide is, *Pray like there's no tomorrow.*

Is political momentum building to take the necessary defensive measures, or is it just 1990 all over again, when (as noted earlier) the North American Electric Reliability Corporation's board of trustees paid lip service to doing the right thing and then just twiddled their thumbs? "In late 2010, a principled senior official from the Federal Energy Regulatory Commission visited the office of Congressman Roscoe Bartlett [R-MD] and told him if the United States were to experience a solar storm of magnitude similar to the 1859 Carrington event, the power grid would collapse for 1–2 years, resulting in the likely death of 75% of the U.S. population," reports Thomas Popik, head of the Foundation for Resilient Societies, a committee of engineers dedicated to protecting technologically advanced societies from natural disasters.[3] Bartlett took this message to heart, and with help from several colleagues, including Congresswoman Yvette Clarke (D-NY) and Congressman Trent Franks (R-AZ), he has indeed worked diligently to champion grid protection. They may possibly succeed, knock wood, in collaborating with the Senate to pass legislation to safeguard the grid from solar disaster. However, judging from the past behavior of NERC and its cohorts, I would be surprised if the utility industry did not do everything possible to stall implementation.

To be fair, there is at least one tough call to make before proceeding with protecting the grid. Regrettably, the surge suppressors that block solar EMP would not be effective in blocking any EMP blasts that might emanate from the detonation of a nuclear weapon in the atmosphere. Retrofitting the entire grid to withstand *all* forms of EMP assault would probably cost in the range of $10 billion to $20 billion, and would likely take a decade or more to implement. Just how likely is it that a terrorist group or rogue nation would manage to acquire not only the requisite nuclear weapon but also a rocket-launched delivery system to explode it in the atmosphere above our country? It would take some pretty sophisticated maneuvering to outwit and outfly our armed forces, who are 24/7 vigilant against such attack. Given the dire budgetary straits in which the federal government finds itself these days, I would have said that we should cheap out and just go for the $1 billion–plus insurance policy against solar EMP rather than the deluxe $10 billion–plus insurance policy against all forms of EMP assault. That is, until I spoke with Curtis Birnbach, president and chief technology officer of Advanced Fusion Systems, a cutting-edge research and development firm. Birnbach is an ingenious inventor who has discovered extraordinary new potential in the seemingly antiquated technology of vacuum tubes. Are you old enough to remember those glass tubes that used to flicker on and off in back of television sets? Well, Birnbach has created high-tech, souped-up versions of those tubes that could be used to protect our society from all forms of EMP assault—even something other than a solar or a nuclear event. What really caught my attention was Birnbach's insistence that EMP weapons need not be nuclear; instead, they could, for example, use x-ray laser technology. Like the EMP-generating device depicted in the recent *Oceans 11* remake, such weapons would be transportable and easily concealed. Moreover, they would not need to be detonated in the atmosphere to be effective; they could do their damage to the grid even when set off from the ground. It is said that a relative handful of these weapons, strategically deployed, could supply enough EMP to knock out many strategic targets in the Boston–Washington corridor.

I am constrained to point out that I have been unable to corroborate independently any of the specifications of these portable, ultrapowerful, nonnuclear EMP weapons. While attending the London summit, I did, on the record, ask Liam Fox, then the British defense secretary, about them. He seemed surprised at the question and would only say that there seemed to be more openness on this subject in the United States than in the United Kingdom. Indeed, the CBS spy-thriller series *NCIS: Los Angeles* recently featured a story (season 3, episode 11) centered on a nonnuclear EMP weapon, demonstrating, if nothing else, that the idea is in the air. Although Birnbach would not say so specifically, I left our conversation with the impression that such weapons are far too advanced for any of our likely terrorist or rogue nation adversaries to develop or perhaps even obtain, though they are probably not beyond the technological reach of Russia and China. Speaking more directly to the credibility of the nonnuclear EMP weapon threat is the fact that Birnbach's firm enjoys the patronage of an illustrious chairman, William H. Joyce, who has served in numerous chief executive capacities atop the chemical industry, and who won the National Medal of Technology, presented by President Clinton, in 1993.

DON'T FRY THE CHIPS

The power grid is not the only thing at risk, by the way. In March 2010, at a conference in Washington state on existential threats, I gave a talk about the dangers of EMP, solar and otherwise, and after my presentation, a pilot for a major commercial airline came up and peppered me with questions. Basically, she wanted to know if such blasts could fry sensitive electronic components used in today's passenger aircraft. Our discussion quickly turned to the case of Air France 447, an Airbus A330-200 that went down en route from Rio to Paris on June 1, 2009, killing all 228 aboard. She had heard the same scuttlebutt I had, that AF447 was done in by solar EMP. When the Airbus went down it was passing through what is known as the South Atlantic anomaly, a California-size crack in Earth's magnetic field—the field that normally protects us from solar EMP. Beginning the day before, on May 31, 2009, and continuing through the crash date, a large sunspot and along with it an enormous plasma filament had erupted on the Sun's northwest quadrant, the quadrant from which Sun-storm explosions are likeliest to hit Earth. Had a blast of solar EMP rocketed through space, penetrated this giant crack in the magnetic field, and fried some of the Airbus's electronic components? Depends on how well those components were protected, or "hardened." Most hardening of electronic components is done under the working assumption that normal conditions will prevail in our planet's protective magnetic field. The South Atlantic anomaly is not "normal conditions"; it is an absence of the normal level of protection.

Ultimately, it was determined that AF447 flew into a terrible thunderstorm, causing a temperature gauge to be coated with ice and thus give false readings, which in turn caused speed sensors to give false velocity readings, ultimately tripping a cascade of failures throughout the plane's computer system—erroneous readings that caused the massive cockpit confusion that ultimately sent AF447 into the ocean. But even the *possibility* that the airplane's electronic components—most of which depend on microchips storing and processing vital information—could have been fried by solar EMP has motivated a flurry of activity in the pilots' union. Understandably, neither the union nor the airline companies want to go into much detail on this publicly. Flying these days is already scary enough.

Suffice it to say that all sorts of microchips, from the ones in airplanes to those in computers and cell phones, road and rail signal machinery, military equipment, electronic car ignitions, microwave ovens, and television sets are considered vulnerable to EMP, regardless of whether such blasts come from the Sun or from weapons. Protecting or "hardening" sensitive electronic components against EMP usually entails completely encasing the devices that contain them in what's known as a Faraday cage, a copper enclosure with a ground. Simple enough, technologically, but cumbersome design-wise—and of course an added expense, though worth it, one would venture, if it keeps your plane in the air.

PRAY FOR A WAKE-UP CALL

Ever have the dentist tell you that a filling that has worked just fine for years has to be preemptively removed because it's going to crack sooner or later and will probably take out a chunk of your tooth with it? Last year I went for that pitch, and my newly refilled tooth cracked a few weeks later! So I can understand why the federal powers that be keep kicking the power grid can down the road. Putting things off is very easy to do, especially when the proposed task involves asking taxpayers to spend money and suffer inconvenience just so that government engineers can tinker around with the electricity supply, which, for the most part, works quite nicely, thank you. "If it ain't broke, don't fix it," as the saying goes, is about as CS (commonsensical) as it gets. No doubt those senators on the energy subcommittee who put the kibosh on the House GRID Act saw the initiative to implant surge suppressors into the grid as an example of federal meddling, yet another slippery slope that would lead to a cascade of burdensome regulations, or worse. Maybe the power grid retrofit will prove more costly and difficult than we anticipated, they probably thought; after all, overruns are the rule, not the exception. Plus, the overhaul could backfire and cause a whole new set of blackouts, brownouts, and such. Oh, how the constituents would yowl about that! Then there's the question of whether we should go for the earlier-discussed deluxe retrofit, protecting against all forms of EMP (including weapons-generated), or settle for the economy plan that protects only against solar blasts. What's the trade-off? Is it like surgery—as long as we're operating on the grid, should we go in there and get everything done at once? Is there a way of doing the overhaul incrementally, starting with one layer of protection and gradually upgrading as the means become available? Lots of questions to be answered here before we pull out the scalpel. Furthermore, the possibility of sabotage cannot be discounted. Whoever gets the contract for EMP-proofing the grid had better be vetted up, down, and sideways. That's all we would need: some closet al-Qaeda operative or homegrown marauder cross-wiring the nation's electrical circuits. Ah, maybe it's best to postpone things for a little while—you know, until the situation becomes clearer. There are, after all, many other urgent demands on our resources and attention.

Given all this uncertainty and confusion, the wait-and-see approach seems balanced and reasonable. In truth, it should work perfectly well—until the lights go out for good.

Procrastination rarely actually *wins* any debate, yet somehow it usually manages to dominate anyway. Think how often "wait and see" trumps the most compelling arguments—in this case, that we must take whatever steps are necessary to protect our electricity supply, that each year we become more and more dependent on the juice (and therefore more gravely vulnerable to the collapse of the system that delivers it). The problem with that argument is that it's *just* an argument. Few will challenge it; most will nod earnestly, then move on to other things. And nothing will be done because . . . nothing will be done.

We have already slept through two wake-up calls from the Sun. On March 13, 1989, a geomagnetic storm knocked out the electric grid powered by Hydro-Quebec. Six million customers went without power for nine hours in the cold Canadian winter. Some very chilly customers did not have their power restored until well into spring. On the day of the event, Minnesota radio listeners found themselves tuned in, not to favorite son Garrison Keillor, but to West Coast voices; for some reason, their radios picked up chatter from the California Highway Patrol. In New Jersey, a $36 million transformer burned up, briefly shutting down a nuclear power plant. Another geomagnetic storm hit on Halloween 2003, knocking out 14 transformers in South Africa and crippling that nation's electricity infrastructure for years. Neither the 1989 nor the 2003 event was unusually large; the 1989 event was calculated to be around 16 percent of the size of the 1859 Carrington event, with the 2003 storm weighing in at even less. Had they been comparable to Carrington, or to either of those subsequent behemoths of 1909 or 1921, your life would likely have borne little resemblance to your existence today, if you existed at all.

In our situation, patience is not a virtue, as the adage would have us believe, but a vice. With each passing year, civilization becomes more and more dependent on the power grid, and therefore more vulnerable to disruptions. Even midsize blasts of solar EMP could short out vast portions of contemporary society. That is a key point here. We ask how the Sun's behavior is changing and what any changes might mean for us. But the simple fact is that our star is not changing nearly as rapidly or as recklessly as we are. Ever since we, as a species, emerged a million or so years ago, *Homo sapiens* has derived most resources locally, supplemented by trade and travel. But over the past century and a half, we have assembled a vast global network of energy and information upon which we have come to depend for our survival. It is as though humankind had grown a collective, technological nervous system external to our individual bodies. This burgeoning global network has enabled us to proliferate in both overall population numbers and average lifespan per individual, securing our ascendancy. With each passing year our population has increased, along with the population of the other species depending upon us for sustenance. Yet this life-giving infrastructure remains exposed and highly vulnerable to disruption, particularly from explosions such as issue from the Sun periodically. A collapse of our technological network right now would likely be less devastating than one that happened 10 years from now, and that would likely be less devastating than one 20 years out. With the clock ticking against us, why have we put so little effort and investment into protecting the grid? We human beings tend to be lazy optimists, gambling that if something hasn't harmed us before, it never will. True, there have been numerous power outages due to equipment malfunctions, weather events, military actions, and even cyber-terrorist attacks. In response, our engineers have added backup systems, though not nearly to the scale of redundancy that would be needed to absorb a repeat of anything approaching the magnitude of the Carrington event.

Is it wrong to hope for a midsize solar blast, one large enough to do some eye-popping damage but small enough so as not to be calamitous? Better a minor crisis now, to my mind, than a major cataclysm down the road. A blast of, say, one-fourth the magnitude of the Carrington event would no doubt burn out a number of transformers, causing inconvenience, perhaps hardship, to millions. The outcry would be such, however, that it would force the utilities to reconstruct the burned-out circuits more safely, hastening the development of our surge suppressor expertise and spurring our resolve to protect the whole power grid once and for all. Huck Finn said, "You can't pray a lie," and I would be lying if I said that I really *wanted* a cautionary solar blast to happen. But I have no problem at all praying that God would do whatever it takes to stir us to the actions necessary to defend ourselves against the wholesale devastation that would result from a widespread grid failure. Is Sol going to have to teach us a lesson but good? My guess is that we're in for the solar equivalent of a slap upside the head. We've already taken a few to the face and have barely learned a thing. Question is, how painful will Sol's next punch be? Enough to leave a mark? Enough to knock us out?

THE ADVANTAGES OF GOING OFF-GRID

Hey, isn't this something that the vaunted new "Smart Grid" is supposed to take care of? Sadly, no. The Smart Grid, an automated, decentralized system designed to monitor, control, and respond to energy supply and demand both nationally and in specific locations, turns out to be even more vulnerable to solar EMP than our current "unintelligent" power network. This is because the Smart Grid relies heavily on SCADA (supervisory control and data acquisition) systems using highly sensitive control devices integrated throughout the network. According to EMPact America, an organization dedicated to advocating for protection of our technological infrastructure from EMP assaults, SCADAs are extremely susceptible to solar blasts.[4] Wouldn't you just know it? The Smart Grid, no matter how complex and artificially intelligent, turns out to be a real dummy.

Given the unreliability of the power grid, whether smart or dumb, is it finally time to spring for those solar panels at home? It couldn't hurt, as my mother likes to say. New England Patriots quarterback Tom Brady and his wife, Giselle Bündchen, Victoria's Secret model turned UN environmental ambassador, have solarized their new $20 million, 22,000-square-foot Brentwood, Los Angeles, mansion, if that helps any. I hope they remembered to shield their inverters, the devices that convert direct current made from the sunlight into alternating current used in the home. This protection can be added for very little extra expense by encasing the inverters in metal boxes that would deflect any potentially harmful solar EMP. The thornier question is how, after the Mother of All Blackouts hit, the duo would cope with having one of the few homes in West Los Angeles that had electricity. They just might receive some unexpected visitors pounding on the door and brandishing appliances. Then again, whatever the downside, it's always better to control a valuable commodity—in this case, electricity—than it is to be in (desperate) need of it. Going solar just might make them a true power couple indeed.

Forsaking the grid entirely, or simply having it as a backup power supply (as appears to be the case with Tom and Gisele's new digs), is frequently held up as a wise way to defend against losing all electrical power in times of emergency. It's a good idea as far as it goes, and might possibly make the difference between surviving a long-term, broad-scale blackout of the type predicted in the earlier-cited National Academy of Sciences space-weather report, and perishing altogether. Solar systems that provide heat and hot water, and that power consumer electronics such as calculators and radios, all stand a better chance of remaining operational after a blackout caused by solar EMP than do their conventional counterparts. So do systems powered by wind, geothermal energy, gasoline/kerosene/propane, and other independent means, including plain old batteries. Whatever takes you off the grid enhances the chances of your household remaining rudimentarily functional. The good news (and bad news) is that you, like Tom and Gisele, will become the envy of all your neighbors, some of whom might not act so neighborly in the midst of the chaos.

Long before most of us get around to going off the grid, though, we may find that the grid has, in a sense, gone off us. In "Why Microgrids Are Inevitable," Peter Asmus makes the case that the electrical power grid is disengaging from itself, as though transforming from a solid mass into an archipelago of discrete yet synergistically connected islands, usually powered by alternative energy sources such as solar and wind, that can be successfully isolated and survive on their own should some catastrophe befall the whole system.[5] Basically, this wave of the future is, for the utility industry, also a return to its roots. The earliest microgrids were invented by Thomas Edison, whose firm had by 1896 installed 58 direct-current versions of them, along with approximately 500 isolated power plants in the United States, Russia, Chile, and Australia. The idea of connecting these microgrids together arose haphazardly, initially as a means of backup: if one city's generator failed, the neighboring generator could be tapped into to help maintain a basic flow of current until the malfunctioning power plant was brought back online. Gradually, the grid came to be used, not just as a safety net, but as the principal supplier of electricity—one into which various generators directly connected their current in order that it might be swapped around efficiently to meet the demands of the broader, interconnected region. The electricity superhighway gradually made microgrids irrelevant, much the way that local main streets and toll roads were bypassed by the interstate highway system.

Asmus defines a "microgrid" as "an integrated energy system consisting of distributed energy resources (DERs) and multiple electrical loads operating as a single, autonomous grid either in parallel to or 'islanded' from the existing utility power grid." Generally, a microgrid is connected to the main power grid at only one juncture, enabling it to be disconnected quickly and efficiently, and without unduly compromising electrical load. Reconstructing the power grid as an interdependent network of microgrids is kind of like rewiring a set of Christmas lights from series, in which the electrical current connects through each bulb socket, to *parallel*, where each socket connects to the trunk line of current. Changing over from series to parallel thus removes the threat that if one bulb goes out, so do all the rest. If individual microgrids are powered by solar, wind, or other renewable energy sources, as seems to be the case with most current microgrid designs, fossil-fuel energy consumption and the greenhouse gas fallout therefrom will be minimized.

So if Tom and Gisele are not among your nearest and dearest, your best bet may well be to connect to a microgrid. Better to be part of a stand-alone community in the midst of the Mother of All Blackouts than just a stand-alone home. I recommend affiliating with the University of California at San Diego (UCSD), my graduate school alma mater. UCSD has one of the most advanced and extensive power supply systems in the world today. With more than 450 buildings metered and sensored from roofs to foundations, and with power generated by solar energy and natural gas supporting such diverse usage schemes as residential dorms, advanced research laboratories, gymnasiums, and mixed-use structures, the 1,200-acre campus could probably survive a Carrington-scale blast to the North American power grid and still function. They had it relatively easy, though: UCSD was designed from the outset to be an integrated whole, making it much easier technologically and logistically to create a coherent microgrid there than, say, in a random 1,200-acre section of San Diego. It didn't hurt that the publicly funded institution of higher learning was crawling with graduate students looking for places to plant their high-tech gizmos. So UCSD shouldn't take too many bows.

On the other hand, Penn South, a moderate-income cooperative residential community of 2,820 apartments in nine buildings in the Chelsea section of Manhattan, has every right to boast, and I don't say that just because my mother lives there. Built by the International Ladies' Garment Workers Union (ILGWU) in 1957, a left-leaning organization if there ever was one, Penn South responded to the soaring energy prices caused by the OPEC oil embargo of the early 1970s by becoming its own microgrid. Residents built on their grounds the Penn South Power House, an independent power plant that essentially obviated the need to connect to Con Edison, the local New York City electrical utility that hands-down holds the record for backlog of consumer complaints. With more than half of Penn South's residents earning less than $40,000 per year and with the average one-bedroom unit currently costing $69,000 (about 10 percent of the price of comparable apartments in the tony neighborhood), this community has very few bells and whistles. But because of its energy independence, it had uninterrupted electrical power during the massive 2003 blackout that darkened much of the Northeast. Come the next Carrington event, there are very few places you would rather be.

QUIETLY ORGANIZING IN THE NAME OF THE SUN

Those with a suspicious turn of mind might do a double take upon learning that a sophisticated data-gathering organization using some of the most technologically advanced instruments in the world has sprung up quietly, with the stated intention of monitoring the Sun–Earth relationship. Since its founding in 2007, the International Space Weather Initiative (ISWI) has grown to include 101 nations providing and/or hosting ground installations for space-weather instruments, sharing real-time data, and organizing seminars, symposia, and educational programs. In an event cosponsored by NASA, the UN Office for Outer Space Affairs in Vienna, the Japan Aerospace Exploration Agency, Kyushu University, and the Bulgarian Academy of Sciences, ISWI convened its first plenary session in Helwan, Egypt, in 2010.

"Strong solar storms can knock out power, disable satellites, and scramble GPS," declared ISWI executive director Joe Davila, of NASA's Goddard Space Flight Center, adding that the Egypt meeting helped the world "prepare for the next big event."[6]

The lion's share of scientific information comes from industrialized nations, most of which are located relatively near the poles. Developing nations, by contrast, are comparatively unexplored in terms of their exposure to space-weather events. What's charming about ISWI is that its annals are filled with reports from far-flung locations, often equatorial nations not known for their scientific endeavors. The typical ISWI report from Nigeria, Bolivia, or Fiji chronicles the installation and deployment of what are known as MAGDAS (magnetic data acquisition system) units that monitor cosmic rays, geomagnetic disturbances, electrical currents in top soil, and auroras, an increasing number of which appear to be flashing down to equatorial regions. Together, these units (once fully deployed and operational) will make up the most comprehensive ground-based monitoring system of Earth's magnetic field and related phenomena. ISWI employment opportunities, though often offered only on a voluntary basis, abound from the high-school level on up. This global push is not just affirmative action, space-weather-style. It is the mining of a new source of vital information about what is known as the *equatorial anomaly*, a "fountain of ionization that circles the globe once a day, always keeping its spout toward the sun. During solar storms, the equatorial anomaly can intensify and shape-shift, bending GPS signals in unexpected ways and making normal radio communications impossible," writes Tony Phillips.[7]

This much we know for sure right now: come the next mammoth, grid-killing solar blast, the equator will be a good place to be. (Remember that EMP effects are felt most strongly near the poles, where Earth's geomagnetic field is weakest.) Sure, the GPS and the radio might get their signals bollixed up for a while, but if the electricity is still flowing anywhere on the planet, it will likely be between the Tropics of Capricorn and Cancer (or thereabouts), including countries in southern Central and northern South America, the Sahara, Micronesia, and South Asia. Highly inconvenient for most of us, though perhaps refreshing in terms of upending the current political and cultural balance of power.

13

A Hundred Nuclear Meltdowns Coming Our Way

No matter how moody the Sun gets, it won't hold a candle to the funk that will settle over the nuclear power industry, and soon thereafter over the rest of us, when the power grid shorts out. Once the electricity goes out, we in the United States will have about a month before the 104 nuclear reactors at 65 sites in 31 states start exploding like so many atom bombs. A pity, because the nuclear power industry has, for the most part, been very well behaved.

Comparably priced, and with lower emissions per kilowatt of greenhouse gases and grimy particulates than its peers—coal and oil—nuclear power is like the model student who throws a beastly

temper tantrum every now and then. Despite its occasional outbursts, the nuclear power industry still places at the head of its
class by any objective cost–benefit analysis of its net impact on
the commonweal. But there's cold comfort in that; it can be hard
to relax completely when you know that there is a lethal tick-tick-
tick counting down somewhere deep inside.

The real problem with nuclear energy is the fear factor, starting
with Hiroshima and Nagasaki, then the Cuban missile crisis, on
up through Three Mile Island, Chernobyl, and Fukushima. The
possibility of a nuclear accident—meltdown!—tends to create a
level of fear far greater than the actual risk to life and limb. Aggravating the situation is the fact that radioactivity is a subatomic
phenomenon few of us grasp beyond knowing that it represents
a cancer risk. In point of fact, significant releases of harmful radiation from nuclear reactors into the surrounding environment
have been quite rare. What's more, the health risks those releases
have posed are trivial compared, for example, to the cumulative
respiratory ailments resulting from the atmospheric effluent of
coal-burning plants. But coal dust is not nearly as scary as radioactivity, so in that case the actual danger outweighs the fear.

To be sure, there are problems with the handling of high-level
radioactive waste produced by nuclear power plants, as well as the
related low-level waste-management issue of how to dismantle
a nuclear power plant, particularly the radioactive components,
once it has been decommissioned. However, within the next
decade or so, these nasty leftovers will be disposed of much
more efficiently, thanks to emerging plasma flame technologies
in which robots handle the waste products from start to finish
onsite, feeding wastes into furnaces that dissociate the gunk completely, leaving no toxic residue. I know this because I was once

chairman of Aerospace Consulting Corporation, an advanced plasma physics research company in Albuquerque, New Mexico, that holds the patent, US 7,026,570 B2, for the Vulcan Plasma Disintegrator, a portable, ultra-high-temperature furnace adaptable for just such a process.

Opponents of nuclear energy have slanted their arguments shamelessly, emphasizing phantom health risks. On balance, the difficulties of securing radioactive wastes produced by nuclear reactors are more than offset by the bounty of clean energy. Moreover, nuclear power plants do not depend on strategically sensitive imports, unlike the oil and natural gas industries, which too often are beholden to treacherous political situations in the Middle East, Venezuela, and elsewhere. That said, it is with a heavy heart that I, a vested proponent of the nuclear energy industry, am forced to warn against its devastating potential.

All commercial nuclear plants depend on an external supply of electricity to function. When the grid collapses—and it eventually will, if safeguards against solar EMP aren't implemented—so will these plants. The shortfall will cause accidents, contaminating large swaths of land with nuclear radiation, thus rendering these areas uninhabitable for hundreds of years. Nuclear power plants are currently designed with the assumption that any disruption in the commercial power supply will be quickly restored. A long-term loss of power of a month or more, such as will likely occur when the next Carrington-size solar blast hits our planet, will lead to meltdown, because the cooling towers will cease to operate. Specifically, a lack of electricity will paralyze the pumps that provide the water to cover and cool the rods that contain spent nuclear fuel, resulting in a meltdown scenario similar to what happened in the Fukushima disaster, where extreme seismic and tidal activity caused the pumps to fail. "If the pumps stop working, the fuel pools will boil, just as the water in a car's engine boils when the water pump fails. If the water in the spent fuel pools boils off and exposes the fuel rods to air, the metal [zirconium] cladding on the rods will catch fire, much like the metal in a 4th of July sparkler. Because spent fuel pools are not in containment vessels, the resulting plumes of radioactive material would be released into the atmosphere," writes Thomas Popik, whose Foundation for Resilient Societies presented a "petition for rulemaking" to the Nuclear Regulatory Commission (NRC) in 2011.[1]

The zirconium fires Popik refers to burn with such rapacity that they are nearly impossible to put out. Under the best of circumstances, massive response logistics would be employed to contain the situation. But the situation I'm describing—long-term loss of power due to a major solar disturbance—is hardly the best of circumstances. Emergency response protocols for preventing nuclear meltdowns generally operate on the assumption that only one plant, two at the most, would be melting down at any given time, meaning that all appropriate resources from around the country and even internationally could be fully deployed to control the situation. However, if the grid were knocked out by solar blasts, in all likelihood many, perhaps dozens of reactors would advance to the critical stage more or less simultaneously, overwhelming our response capability severely, since the rescuers themselves would most likely be without power as well.

This fatal vulnerability of our nuclear power infrastructure has been amply confirmed by high-level technical reports, including the series *Electromagnetic Pulse: Effects on the U.S. Power Grid*, from Oak Ridge National Laboratories, prepared in 2010 for the

Federal Energy Regulatory Commission (FERC) in conjunction with the Departments of Energy and Homeland Security.
The Oak Ridge reports indicate that most of the United States
east of the Mississippi River as well as the northwestern states of
Washington, Oregon, and Idaho are the likeliest areas to suffer
grid collapse as the result of a solar storm.[2] That's unfortunate,
because 71 out of the total 104 nuclear power plants are within
these red zones. More than two million Americans live within
10 miles of nuclear power plants located in these high-risk areas.
Evacuation procedures typically count on the fact that telephones
and emergency sirens would be working, probably not the case if
the electricity were knocked out. Popik's group advocates requir-
ing nuclear power plants to install backup cooling systems that
operate independently of the grid. The estimated cost for such
systems would be paltry, around $150,000 each.

The NRC appears to be taking this all "under advisement." In
these challenging economic times, any governmentally imposed
expenditures meet with fierce resistance, and such proposals often
end up being political footballs in the never-ending free-market
debate about regulations. Installing backup cooling pools would
add another layer of complexity to what is already a profoundly
intricate energy production and delivery system, and would also
rouse watchdogs on the lookout for any new evidence that could be
used against the nuclear power industry. All that said, Popik's plan
seems sound, just so long as it is not seen as a substitute for protecting the grid, whose collapse would cause the problems in the first
place. The ultimate solution must come from the power grid itself,
over which the NRC has no direct authority.

The good news is that nuclear contamination is just one small
element of the power grid collapse scenario. This is also the bad

news. Compared to the rampant chaos that would result from months or years of little or no electricity (and therefore little or nothing in the way of the basic services that depend on it), even a chain of meltdowns would pale in comparison to the toll taken by general societal upheaval. This is not just a simple calculation of the loss of life and property, though that loss would be horrific. We must also consider the psychological dimension of internecine madness. People have an enormous ability to take calamity in stride, especially when, as in the case of an accidental nuclear catastrophe, there is nothing personal about that calamity. However, when the suffering comes at the hands of others—such as might well be the case when scarcity of food, clean water, and medicine becomes intolerable and leads to bloody violence—a tit-for-tat hunger for revenge will pass from one generation to the next on down the line, staining and deforming our future. But ask people which scenario they fear more, nuclear catastrophe or social upheaval, and I'll bet that nuclear catastrophe is deemed scarier. Once again, the nuclear power industry is a lightning rod for our anxiety. However unjust, that does seem to be its predestined role in the contemporary psyche. After all, its grandparents are Hiroshima and Nagasaki. We are used to being afraid of things nuclear, and we are okay with that.

14

The Sun Will Send Us Secret Warnings

Contemplating such immense possibilities as the destruction of society can lead to startling observations. Take, for example, an oft-quoted remark from Admiral Hyman Rickover, who oversaw the transition to a nuclear U.S. Navy. He said that we ought not worry, should nuclear war wipe out humankind, because in that case a new and wiser species will evolve. Equally macabre, though much more hopeful in the short term, is recent research indicating that the Sun has been sending vitally important messages to piles of radioactive medical waste. I kid you not.

One day in early 2011, I received an e-mail that said, in full: "I am just wondering how you knew this?" It seems that someone had heard me interviewed about my visit to Akademgorodok, the earlier-mentioned sequestered scientific research city in Siberia,

where I learned of these secret messages, an utterly strange phenomenon that U.S. researchers apparently thought was known
only to them. Embedded in that e-mail was a link to an article
describing how physics and engineering professors from Stanford
and Purdue had come to believe that the Sun sends secret messages to Earth, and perhaps also the other planets, that could tip
us off as to when the next big solar blast is going to hit. It turns
out that certain radioactive elements here on our planet begin to
decay more slowly 24 to 48 hours before a flare erupts on the Sun
and then travels toward and hits Earth.

"On December 13, 2006, the sun itself provided a crucial clue,
when a solar flare sent out a stream of particles and radiation
toward Earth. Purdue nuclear engineer Jere Jenkins, while measuring the decay rate of manganese-54, a short-lived isotope used
in medical diagnostics, noticed that the rate dropped slightly
during the flare, a decrease that started a day and a half *before* the
flare," reports Dan Stober in *Stanford University News*.[1]

SOLAR STORM BAROMETERS

Just as hurricanes broadcast warnings in the form of low baromet-
ric pressure days before they begin to whip up, when these radioisotope barometers start falling it means that a megastorm could
be on its way from the Sun. "If this apparent relationship between
flares and decay rates holds true, it could lead to a method of predicting solar flares, prior to their occurrence, which could help
prevent damage to satellites and electric grids, as well as save the
lives of astronauts in space," writes Stober.

How could this be? The best guess at this point is a seeming
impossibility: solar neutrinos, particles that have no charge and

no mass while at rest and only infinitesimal amounts even when traveling at near light speed, somehow interfere with the clockwork regularity of radioactive particle emissions 93 million miles away. What's more, some of the decay-rate aberrations occur in the middle of the night, meaning that the solar neutrinos— deemed incapable of following anything but a straight path—had sailed all the way through Earth to do their thing. "No one knows how neutrinos could interact with radioactive materials to change their rate of decay. . . . It doesn't make sense according to conventional ideas," says Ephraim Fishbach of Purdue. Jere Jenkins adds, "What we're suggesting is that *something that doesn't really interact with anything is changing something that can't be changed*" (italics mine).

So phantom zeroes a zillionth the size of the period at the end of this sentence have been tipping us off to the fact that solar blasts are on the way, eh? If the mystery particle is not a neutrino, Stanford's Peter Sturrock observes, "It would have to be something we don't know about, an unknown particle that is also emitted by the sun and has this effect, and that would be even more remarkable." Regardless of not knowing precisely how this process works, "All of the evidence points toward a conclusion that the sun is 'communicating' with radioactive isotopes on Earth," concludes Fishbach.

Does the Sun have a *tell*, an unwitting giveaway of its stormy intentions? Or is there some purpose behind the warnings? It seems likely that these cryptic solar missives, whether deliberate or unwitting, will serve as advance notice of impending geomagnetic storms. To determine if this "secret message" phenomenon is reliable as a forecasting tool, meticulous note will have to be taken of every solar blast that hits Earth, and the response, or lack thereof, in decay rates of the radioisotopes. This could one day save greatly on expenses associated with solar research satellites. NASA, ESA (European Space Agency), and collaborating space organizations around the world will not be thrilled if flashy and expensive IHY projects zooming around the solar system are trumped by meters stuck in vats of radioactive goo back home! Think of all the money that could be saved, though. Right now, state-of-the-art technology in the early detection of sunspots involves a complex and expensive forensic tool that recognizes the sound waves made by sunspots while they are still buried beneath the Sun's surface. It's a (very) long-distance version of Cold War–era spy gear that could record the vibrations of windowpanes and from those data reconstruct the conversations that made the glass shake in the first place. Difference is that we don't speak "sunspot" and thus far have not been able to determine which of the spots bubbling to the surface will make explosions that will actually hit our planet. That's the beauty of the Stanford–Purdue technique. Those radioisotope detectors will register only when rogue neutrinos are streaming from the Sun to Earth, presaging a full-scale earthbound blast.

Forewarned is not necessarily forearmed, however. Unless utility operators are willing to place enough faith in radioisotope decay data to shut down major portions of the power grid preemptively, the warnings will prove to be of very limited value. To be fair, it is a lot to ask a utility operator—to cut off power to commercial, governmental, and residential customers, thereby risking not only loss of income but a welter of lawsuits, particularly if the predicted threat never materializes. It is one thing to justify a multi-billion-dollar decision based on research data collected from government spacecraft monitoring the Sun's behavior, quite another to base it on an arcane, inexplicable process that its discoverers would until very recently have considered a physical impossibility. Of course, the game would change dramatically if the utility industry ignored an accurate prediction made by an analysis of the behavior of those rogue neutrinos and then suffered a big, grid-paralyzing solar blast. But that would be a very large price to pay. Bottom line: we still need to install surge suppressors to protect our electricity supply.

THE SUN AS BROADCASTER

Warnings are being sent from outer space to piles of radioactive waste? What a fantastic concept! Who knows where it will lead? The researchers from Stanford and Purdue have, in courtroom parlance, opened the door to all sorts of speculation. Could the Sun be sending us other secret messages? There's sure to be a research scramble to discover more such clandestine forms of Sun–Earth communication. Imagine a new vision of the Sun as broadcasting hundreds, even thousands, of as yet undiscovered forms of energy and information to our planet. How godlike will this fearsome star become in our eyes?

The Sun could also be jamming our wavelengths. What if those rogue neutrinos have been messing with carbon-14 and beryllium-10? As noted earlier, so much modern science—practically all of our assumptions about Earth's history before, say, a century ago—depends on the premise that carbon-14 and beryllium-10 radioisotopes decay at a steady rate. As of this writing, there is no evidence to indicate otherwise, although the research on this decidedly weird solar neutrino phenomenon is still in its infancy. The Stanford–Purdue findings focus on radioactive manganese and cesium, indicating that the decay rates of these isotopes slow very slightly due to the above-described effect. Even if the carbon-14 and beryllium-10 decay processes prove to be susceptible, the slowing effects would likely be just as minimal, and would therefore not mitigate the isotopes' usefulness in ascertaining the age of geological samples such as ice cores and tree rings. Or so the scientific establishment will vociferously maintain. Too many careers, grant millions, and prestigious prizes are at stake. No one, including the author of this book, wants to have to say, "Oops! We have to completely rethink Earth's history. Sorry about that," because of some weirdo rogue neutrinos or other solar voodoo.

Unlikely as it seems, though, just as the Sun sends secret warnings to piles of radioactive waste, so too it may well be transmitting other types of messages that we have not yet figured out how to decode. Quite likely, we will soon learn myriad and astounding things about the messages our star is sending us. The Sun–Earth relationship has been moving into the research spotlight as a result of the massive scientific collaboration connected with the International Heliophysical Year 2007–2008. Expect a wave of interest in how variations in solar behavior affects, well, pretty much everything. The Sun is "in" these days, and trendiness matters to the grant-givers. One promising line of inquiry regards Earth's protective magnetic field, which, as explored in the next chapter, has recently shown signs of deteriorating. Diminished magnetic field protection might well intensify the solar messaging effect.

15

Three Looming Threats and One Happy Ending

Compounding the potential dangers discussed in previous chapters—dangers that Earth would confront in the event of a major solar disruption—are several specific threats that make those dangers more likely: deterioration of Earth's magnetic field, Earth's shrinking proximity to a turbulent interstellar energy cloud, and variations in Earth's "wobble."

A DETERIORATING MAGNETIC FIELD

Our planet is a giant electromagnet, with force emanating from the rotation of its molten iron-nickel core. Just as in any other magnet, Earth's lines of force are emitted from one pole, traverse its length, and then reenter at the other pole. Periodically, our magnetic field undergoes a "pole shift," or switch in polarity. Such shifts do not entail any physical movement of land masses, but rather a long, slow flip-flop in the poles' north/south functioning—with magnetic poles showing up at intermediate locations around the globe during the transition. After a shift, the magnetic pole that was formerly emitting lines of force now absorbs them, and the pole formerly absorbing the lines of force now emits them. (The North Pole is traditionally defined as the pole from which force emanates and the South Pole as the one where force reenters. However, there have been so many pole shifts in geological history that the "north" and "south" designations no longer line up with geographical north and south.) Earth is believed to have undergone between 5,000 and 10,000 such pole shifts since it was created 4.57 billion years ago. This belief is based on research of the planet's paleomagnetic record. Studies of the walls of ancient cliffs whose layers correspond with geological eras (the deeper down the layer, the older it is) reveal that every half million to one million years or so, there is a 180-degree shift in the direction in which magnetic materials point. Had there been no pole shifts, magnets throughout all layers would be lined up in the same north–south direction.

The most recent pole shift occurred around 740,000 years ago and is believed to have taken two or three millennia to complete, during which time the planet's overall magnetic field weakened significantly. There is nothing inherently dysfunctional about this process; indeed, periodic pole reversals may somehow ultimately help refresh the integrity of Earth's magnetic field. From Gaia's perspective, letting down her guard every now and then might prove rather exhilarating: she is pierced by Sol's rays while her magnetic cloak is askew. Sol, by contrast, switches poles all the time, every 22 years or so, at every other solar climax. This works out to about 100,000 shifts per Sun year—that is to say, the amount of time it takes for the Sun to complete a single revolution around the center of the Milky Way galaxy. One Sun year works out to about 225 million Earth years. The next solar pole shift is anticipated to commence in late 2012 and reach completion by the end of 2013. During a solar pole shift, storm activity in the form of sunspots and other disturbances reaches a peak. During periods when the solar and terrestrial pole shifts coincide, our planet is maximally exposed to plasma bursts and other random energy inputs from the Sun and cosmos.

Like Gypsy Rose Lee, the legendary burlesque queen, Gaia seems ready to strip once again, except that instead of dropping clothes she is dropping her magnetic cloak. Although the process might seem painfully slow even to Gypsy Rose, who could take 10 minutes to remove a single glove, in the utterly ancient lifespan terms of Gaia, even a full-on, 3,000-year pole shift would be like doing an entire striptease in 15 or 20 minutes. As I examined in *Aftermath*, there are growing indications that our planet's magnetic field has destabilized. The north and south magnetic poles have been moving an average of 40 kilometers (25 miles) annually, with the North Pole apparently bent on exiting North America for Siberia. This shifting is beginning to have some practical consequences. For example, the wandering of the poles has forced a number of airports, including Tampa International, to recalibrate their orientation. (Runways are named numerically, based on degrees of orientation toward compass points; as the poles shift, so does a runway's degree of orientation. Eventually, that discrepancy becomes great enough to matter from a navigational point of view.)

Normally, Gaia's magnetic field prevents solar plasma radiation from hitting the planet's surface, as I briefly noted earlier. It does this by sending that radiation into high orbital patterns known as the Van Allen radiation belts. Major solar blasts overfill these belts, with the leakage spilling out to form the majestic northern and southern lights known as auroras. But these are not normal times. In mid-December 2008, an unbelievably large, pole-to-equator hole in Earth's magnetic field was discovered when a squadron of five NASA solar research satellites, collectively called THEMIS, flew unexpectedly through the enormous breach, their findings standing astrophysics on its head. "When I tell my colleagues, most react with skepticism, as if I'm trying to convince them that the sun rises in the west," says THEMIS project scientist David Sibeck of the Goddard Space Flight Center. "This completely overturns our understanding of things. . . . This could result in stronger geomagnetic storms than we have seen in many years."[1]

The shields are down, Scotty. And there's no telling when they will come back up. For Sol and Gaia, it's just business as usual. But for us, the thinning of our shield against solar and cosmic radiation is tantamount to stripping the skin off our bodies or the insulation off the wires that connect the electrical power grid. Heightened exposure to solar and cosmic radiation resulting from diminished magnetic protection will also likely increase the incidence of mutations, and therefore cancers, in terrestrial organisms.

Cold comfort that pole shifts may turn out to be pivotal for our long-term evolutionary development. The greater the incidence of mutations, the faster evolution progresses: that's the rule of thumb. So how might the biosphere evolve differently if Earth's magnetic field weakens significantly? It could be that pole shifts contribute to the development of superior navigational skills; Earth's skin veritably crawls with creatures making their way from one place to another. During a pole shift, both the incremental changes in navigational degree and the fact that multiple poles may pop up around the globe will create problems, because humans, animals, and machines all count on the traditional two-pole magnetic field structure as a frame of reference. Imagine a world where compasses point in different directions depending on where they are located, pointing south in London, east in New York, and west in Beijing. Chaos! Whether in Boy Scouts' backpacks, on airplane instrument panels, or embedded in brains, all compasses would be susceptible to this confusion. No one would be able to locate, find their way to, or home in on any point reliably—except for those superior ones who develop a better sense of direction.

THE TURBULENCE OF AN INTERSTELLAR ENERGY CLOUD

God forbid that Sol and his planets lose their way in interstellar space and end up getting masticated like so many sunflower seeds, as happened recently in the constellation Draco. On March 28, 2011, instruments on the orbiting spacecraft serving NASA's Swift mission—an internationally sponsored mission devoted to the study of gamma-ray bursts—detected an explosion as bright as 100 billion suns. It occurred, the instruments reported, some 3.8 billion light-years away. Astronomers believe that this gamma-ray burst was the death rattle of a star (apparently about the size of our Sun) as it was devoured by a huge black hole, a hole one million times its victim's mass, at the center of an un-catalogued galaxy referred to simply as Redshift 0.3534. "This is truly different from any explosive event we have ever seen before," said Joshua Bloom, the UC Berkeley astronomer who first noted the significance of the event.[2]

Gamma rays make x-rays look fat and lethargic. The ultratiny, ultra-high-frequency gammas, smaller than the diameter of the nucleus of a typical atom, are natural cancer causers because they can so easily penetrate and fracture DNA. Fortunately, Earth's atmosphere usually absorbs most incoming cosmic gamma radiation, but there was nothing usual about what went down in Redshift 0.3534. Bloom believes that the carcass of this particular star heated up as it was dragged toward the black hole, then swirled as though going down a drain, screaming out torrent upon torrent of gamma radiation as it was sucked in and dismembered. Usually such bursts peter out in a few days, maximum, but in this case the radiation jet stream lasted for months, and Earth just happened to be in the bull's-eye.

Black holes are at the center of most galaxies, including the Milky Way. If a star randomly wanders too close to one of these dark suckers, it is shredded, along with any and all planets and hangers-on. Fortunately, the odds are vanishingly small that such carnage as occurred in Redshift 0.3534 will happen here anytime soon; astronomers handicap it at about once every 100 million years in any given galaxy. This means Sol and Gaia have dodged such gruesome cataclysms 40 or 50 times since their relationship began.

Though it will never amount to anything approaching the

magnitude of Redshift 0.3534, there is a situation at the edge
of the solar system that may well negatively impinge on life on
Earth within our lifetime. Like the wealthy Wall Streeter who
takes a wrong turn into a rough Bronx ghetto in Tom Wolfe's
novel *Bonfire of the Vanities,* our solar system appears to have
blundered into a bad neighborhood. As a matter of fact, we
appear to be headed into a bonfire of cosmic proportions. Even
though we might know it intellectually, many of us find it difficult to picture the solar system as actually on the move. If we
do grasp that fact, it's usually with the assumption that the Sun
and planets are hurtling uneventfully through the starry void.
That our system could actually crash into something seems quite
implausible, at least on any timescale relevant to our personal existence. The truth of the matter, though, is that Sol forever leads
us on a merry chase—well, not always so merry.

The startling hypothesis that our solar system, including the
Sun and all its planets, asteroids, and comets, is moving into a
potentially dangerous and destabilizing interstellar energy cloud
has been resoundingly sustained. "A strong, highly-tilted interstellar magnetic field near the solar system" has been described
in *Nature,* a deeply respected scientific journal, reporting on
data transmitted from the Voyagers, twin space probes that have
been exploring the outer reaches of the solar system since 1977.[3]
The findings were published on December 24, 2009. Merry
Christmas!

"We have discovered a strong magnetic field just outside the
solar system. This magnetic field holds the interstellar energy
cloud together and solves the long-standing puzzle of how it can
exist at all," says lead author Merav Opher, a NASA guest inves-
tigator, currently of Boston University. Opher explains that this

energy cloud is at least twice as strong as had previously been predicted and that the solar system has begun to pass into it, adding that this field "is turbulent or has a distortion in the solar vicinity."

We blindly revere Sol as the sovereign power of the solar system, never stopping to think that the star that guides and protects us as we travel through interstellar space might actually be kind of puny relative to its galactic surroundings. We tacitly assume that our space weather will never turn inclement, and therefore take for granted that our current level of preparedness will suffice indefinitely. In point of fact, however, local space weather has already begun to worsen.

"Local Fluff" is what astronomers call the cloud we are currently bumping into. It's about 30 light-years (180 trillion miles) wide, about 6,000 degrees Celsius, and wispier than an adolescent's Fu Manchu. Nonetheless, the friction from passing through such an energy cloud would be enormous. The Fluff wraps itself around the leading edge of the solar system and heats everything up. That process, called the bow shock effect, is analogous to what happens when the nose of a space shuttle superheats when reentering Earth's atmosphere.

Voyager data indicate that the Fluff is itself surrounded on three sides by an enormous bubble of million-degree gas left over on from the explosion 10 million years ago of a nearby supernova cluster. (Recall that the fourth side of the Fluff is flush against our solar system.) So how does a low-density, comparatively cool entity such as the Fluff avoid being consumed by the million-degree supernova gas? It has an exceptionally powerful magnetic field holding it together. "Voyager data show that the Fluff is much more strongly magnetized than anyone had previously suspected. . . . This field can provide the extra pressure required to resist destruction," writes Opher.

From our solar system's perspective, it's quite a good thing that the Fluff's magnetic field is strong enough to keep it from being crushed—something to thank God for, if one is so inclined. For the Fluff is what stands between us and the million-degree supernova gas. Without it, we would all be dissociated into the basic elements that constitute our bodies and, perhaps, souls. May the good Lord and/or dumb luck continue to protect us as we pass through the Fluff, which, it must be noted, at 6,000 degrees is not nearly as hot as the supernova cloud, but still plenty hot enough on its own to flame-broil us but good.

Up until now, the Fluff has been barred from entering solar system space by a frothy sheath of magnetic bubbles, each one about 100 million miles in diameter, or so the Opher team reports: "The sun's magnetic field extends all the way to the edge of the solar system. Because as the sun spins, its magnetic field becomes twisted and wrinkled, a bit like a ballerina's skirt. Far, far away from the sun, where the Voyagers are now, the folds of the skirt bunch up," says Opher. In other words, our physical reality is being protected by the cosmic equivalent of a tutu!

To get what's going on at the edge of our solar system, picture three balloons bunching up hard against each other. Balloon #1, otherwise known as the heliosphere, protects the Sun and all the planets from intruding energy clouds, galactic cosmic rays, and myriad other assaults and disruptions. The Sun created the heliosphere and keeps it inflated with solar wind. Balloon #2 is the Fluff, a medium-hot, low-density mixture of hydrogen and helium that is pressing up against Balloon #1. Balloon #3 is the supernova balloon, a superheated gas bag that is trying to engulf and incinerate the Fluff. What are the odds of at least one of these balloons popping? Opher and colleagues do not speculate. Neither do they examine the earthly ramifications of our solar system converging with the Fluff, beyond suggesting that we could face an increase in cosmic rays. This, as explored elsewhere in this book, would affect everything from space travel to rainfall.

Much less reticent to connect the dots is Alexei Dmitriev, an esteemed Russian space scientist who has been publishing on the subject for the past 15 years. As detailed in my recent book *Apocalypse 2012*, I visited Dmitriev in Akademgorodok, Siberia. Based on his team's analysis of Voyager data, Dmitriev observes that the atmospheres of Jupiter, Saturn, Uranus, and Neptune are inexplicably excited, as evidenced by immense storms, mammoth eruptions, and plasma arcs jetting from the planets' surface to their moons. He reasons that this turbulence is caused by an external injection of energy into the planets' respective atmospheres. This energy comes from an interstellar energy cloud—the Fluff that we've been discussing—which the leading edge of the solar system has now entered. Dmitriev observes that passage into the Fluff has already begun to perturb the Sun, causing solar outbursts that are leading to hurricanes, earthquakes, and volcanoes of unprecedented ferocity here on Earth. He is on record as predicting that we will face global catastrophe in "not tens but ones of years." When pressed, Dmitriev guesstimates that the solar system will remain within this turbulent milieu for something on the order of three millennia.[4]

Strange that news about the interstellar energy cloud hasn't gotten more play, given that distinguished researchers from both the United States and Russia have warned of it. The findings may be perceived as being of too great a magnitude, potentially causing people to go tilt; could be the powers that be want the whole thing downplayed to prevent panic. Then again, why worry about something you cannot possibly control? The only thing to do is kick back and pray.

VARIATIONS IN EARTH'S WOBBLE

"Keep your eyes wide open before marriage, half shut afterwards." Gaia might do well to follow the second part of Benjamin Franklin's hoary injunction. Scuttlebutt has it that Sol has a babe stashed just outside the solar system. Nemesis, as this secret lover has come to be known, is said to average about one and a half light-years (9 trillion miles) away, closer, by far, than any other star. She's not very luminous, which is why we can't see her even though she is so close. If Gaia were the jealous type, she might harp on the way Sol lavishes time and attention on his molten mistress. Sol and Nemesis orbit each other; there's an invisible central point of gravitational balance between them. By contrast, Gaia does all the revolving; relative to her, Sol never moves an inch.

While the notion of our sustaining star having a paramour might scandalize some, others simply shrug philosophically, à la Ben Franklin. After all, astronomers guesstimate that 55 percent of the stars in our Milky Way galaxy have orbital companions, making the odds slightly better than half that Sol would have one too. The Nemesis ("dark star") Hypothesis was proposed in 1984 by Richard A. Muller,[5] the same UC Berkeley astrophysicist who much later in his career "broke" the global warming hockey stick.

Muller had trained as a graduate student under Nobel physics laureate Luis Alvarez, best known for his "impact" theory that the dinosaurs, along with 70 percent of other species then living, were extinguished by the impact of a comet or asteroid. Alvarez and his son Walter discovered a layer of iridium, an element otherwise found only in meteorites and other extraterrestrial objects, some 62 million years "down" in the fossil record, the same timeframe in which carbon-14 dating of dinosaur bones shows the giant reptiles (and others) to have perished. The subsequent discovery off the coast of the Yucatán Peninsula of an impact crater dating back to that cataclysmic time, now known as the K-T (Cretaceous-Tertiary) boundary—that is the boundary between those two geological periods—pretty much sealed the deal.

Alvarez challenged Muller to find out if this mass extinction was a unique event, or just one of a series of related cataclysms that happened periodically. Evidence had piled up indicating that terrible mass extinctions had indeed occurred before the one that took out the dinosaurs, though no one had demonstrated how any of these calamities could be related to each other. Muller responded to Alvarez's challenge with the Nemesis Hypothesis, which postulates that every 26 million years or so, the dark star's orbit comes too close, setting off a gravity tsunami with catastrophic consequences for the solar system, including our Earth. Intriguing as the Nemesis Hypothesis is, Muller could never really produce any corroborating evidence. Eventually his attention shifted to an in-depth study of the fossil record, where he found that something causes mass extinctions about every 62 million years, precisely the amount of time, it so happens, since the one that occurred at the K-T boundary. Muller's calculations of the periodicity of such megacataclysms have been independently verified in *Nature* to a 99 percent statistical certainty, meaning that the next ultraviolent mass extinction episode could happen at any time, and in fact may be slightly overdue.[6]

Most of the investigation into Nemesis now has to do with Earth's wobble. Our planet wobbles on its axis, but why? Orbital vagaries of Sun and Moon are the usual explanations, vacillations in their gravitational fields affecting our orbit. But that just *shifts* the question rather than answering it. What, then, causes the solar and lunar orbits to behave so weirdly? The answer, according to Walter Cruttenden of the Binary Research Institute, which is dedicated to proving the existence of the companion star, is that Nemesis is shaking things up from behind the scenes, just as an illicit lover might do. Cruttenden holds that the dark star's orbital flybys occur once every 25,200 years, roughly a thousand times more frequently than Muller had thought.[7]

This figure is based on an analysis of Earth's orbital wobble, which, whatever its cause, is widely believed to underlie the phenomenon known as the *precession of the equinoxes*, the process by which stars are perceived to shift, in relation to Earth, about one degree every 70 years. Think of it this way: if you looked up at the same star from the same location on the same date every year, after 70 years that star would seem to have moved one degree, about the width of your thumb. If our planet were perfectly wobble-free, the stars would not "shift" their positions at all. Multiply 70 years by 360 degrees and you get 25,200 years for the entire sky to come full circle, which is why most Nemesis proponents, including Cruttenden, now sign on to that figure as the length of the dark star's orbit around Sol.

Sound farfetched? The Nemesis Hypothesis was on its way to being relegated to the ash heap of science, what with Muller walking away from the debate and no one else able to come up with anything conclusive. But just as Sol was about to be acquitted of the charge of having a deranged lover—think *Fatal Attraction* on a cosmic scale—new and damning evidence emerged. Most important was the peculiar 10,000- to 15,000-year orbit of Sedna, a planetoid discovered in 2006 by Michael Brown of the California Institute of Technology. Brown, no one's idea of a fringe scientist, is best known for having advanced the arguments that removed Pluto from our solar system's pantheon of planets.

"Sedna shouldn't be there," notes the Caltech astrophysicist. "There's no way to put Sedna where it is. It never comes close enough to be affected by the sun, but it never goes far away enough from the sun to be affected by other stars. . . . Sedna is stuck, frozen in place; there's no way to move it, basically there's no way to put it there—unless it formed there. But it's in a very elliptical orbit like that. It simply can't be there. There's no possible way—except it is. So how, then?" Brown admits that his research methodology, which focused on small, speedy objects, would have missed something "so large and slow-moving as Nemesis," an object massive enough to explain Sedna's loopy orbit.[8]

So what if some obscure planetoid a gazillion miles away isn't precisely where expected? Brown has not speculated publicly as to whether or not Nemesis is a threat, but on that subject, Cruttenden is like a kid in a candy shop. He swears that Nemesis periodically triggers gravity tsunamis that profoundly impact our solar system, causing mass extinctions on Earth. The implications for the rise and fall of human civilization are profound. For the sake of our family, Sol should dump Nemesis, though of course he cannot. The two stars are slaves of gravity, trapped in a lifeless dance. Like the lovers whirling endlessly around the second circle of Dante's hell, they are condemned to each other for the rest of eternity. Let us hope that Nemesis, should she exist, never wanders too near to us, nor does anything that would put her lover Sol in a foul mood.

SOL AND GAIA TOGETHER AGAIN

"Reunited and it feels so good."

It seems unlikely that Peaches and Herb, who sang "Reunited," the above-quoted triple-platinum *Billboard* hit, recorded their love song with the reunion of Sol and Gaia in mind. Nonetheless, it would be the perfect anthem to play on December 19, 2024, when Sun and Earth will be physically reunited for the first time in 45 million centuries.

Gaia is the one making the move. As noted in the introduction to this book, a score or so of spacecraft have been sent from our planet to study our star over the past half century. Typically, these research satellites collect their data from solar orbit. But if all goes according to schedule, and assuming that the remainder of its megabucks funding does not get axed, Solar Probe Plus, composed of two automobile-size spacecraft, after its launch in 2015 will come far closer to the Sun than any previous probe—about 3.7 million miles from the surface—and will actually, physically pierce the Sun's corona, essentially the Sun's atmosphere, to analyze its plasma.

The corona is perhaps the greatest enigma in solar physics. At one to three million degrees Celsius, it is hundreds of times hotter than the Sun's visible surface four million miles closer to its core. How weird! The heat generated in the core, the hottest part of the Sun, dissipates as it reaches the surface and then somehow reintensifies in the corona? Solar physics has long had it that our star has an extraordinarily hot core powered by thermonuclear fusion, the same force that powers H-bombs. This vision of how the Sun works was advanced in 1926 by British astronomer Sir Arthur Eddington, who believed that the Sun is essentially a giant ball of plasma—energy in gaseous form—powered by thermonuclear fusion explosions radiating heat and light outward from the core. Eddington's explanation has the great advantage of explaining why the Sun's gravity doesn't cause it to implode and become a black hole or something similar. "It is not enough to provide for the external radiation of the star. We must provide for the maintenance of the high internal temperature, without which the star would collapse," writes Eddington.[9]

Then again, why should gravity cause the Sun to collapse? It doesn't cause the air to collapse, or fire. Eddington's critics argue that he set up a false hypothetical condition to bolster his own theory. Yet his hyperhot core model has become so entrenched that few solar physicists deign to consider any alternatives, despite an array of contrary evidence. For example, common sense dictates that the farther out one goes from the star's thermonuclear heart, the cooler the Sun should get. The problem, as noted, is that the Sun gets *hotter* as one goes out from the surface. Moreover, under Eddington's model, the Sun's energy should dissipate into formlessness as it reaches farther and farther out into space. Instead, solar output seems contained in the heliosphere, an envelope with distinct boundaries and characteristics. Even Eddington himself acknowledged the contradictions, more openly and graciously, one might add, than most of his contemporary disciples: "It must be confessed that [this] hypothesis shows little disposition to accommodate itself to the detailed requirements of observation, and a critic might count up a large number of 'fatal' objections," Eddington wrote.

To reconcile these glaring contradictions between sacred theory and observed fact, a veritable Cirque du Soleil of intellectual handsprings have been performed. Two of the most popular: 1) conveyor belts whisk the heat out from the surface to the corona; 2) dimly understood gravitational anomalies invert the heat convection process. The common thread among these and other sometimes tortured explanations is to regard the Sun as a great, boiling cauldron in which energy works its way up and out as heat. How might that process work? One school stresses that the boiling process is not only hot but very loud. Its proponents believe that sound waves from within Sol's noisy, boiling body work their way up to the corona and turn into heat. Others amend that theory somewhat to say that instead of sound becoming heat, its magnetism—magnetohydrodynamic waves, to be precise—percolates its way up through Sol's intricate magnetic infrastructure to the corona and becomes heat. A third school believes that electric currents work their way through the Sun's internal magnetic field, become hopelessly tangled up, and then explode like a mass of short circuits, releasing bundles of heat.

And then there are the renegades who just don't buy any of those "up and out" heat dissemination scenarios. They wonder if the Sun isn't more complicated than that, that it not only generates heat from its core but also *receives* energy like an antenna. The idea is that dark matter, thought to constitute almost one-quarter of the universe, is somehow captured by the corona and converted into heat energy. Dark matter is a useful construct to explain why stars and galaxies have stronger gravitational fields than their apparent mass would indicate. Imagine a dozen bags of potatoes identical in every way, except that six of them weigh 10 pounds apiece and six weigh 12.5 pounds apiece. To explain the discrepancy, one might posit the existence of an invisible substance that somehow inhabits the heavier potato bags. This is the idea behind dark matter. The whole universe weighs more—has stronger gravitational attributes—than scientists calculate it should. The numbers just don't add up. Thus, the Dark Matter Hypothesis, first advanced by Fritz Zwicky of the California Institute of Technology in 1933. The problem with explaining the corona enigma by means of dark matter combustion is that the proposed (anti)substance is virtually undetectable except by inference, meaning that its properties are largely unknown.

Solar Probe Plus will be our best bet ever for solving the Sun's mysteries. It is expected to launch in 2015, as noted earlier, but will not make close contact with the Sun for almost a decade—thus the December 19, 2024, proposed date for Sol and Gaia's reunion—as its orbit methodically zeroes in on its superhot target. Dick Fisher, director of NASA's Heliophysics Division in Washington, D.C., insists that the probe is designed to resist the corona's heat and to function properly at temperatures up to 1,500 degrees Celsius.[10] (Blisteringly hot though this is, such temperatures are only 1/1,000th of the corona's maximum, meaning that only its coolest, outermost portion can be directly sampled by the spacecraft.) The probe will conduct numerous experiments examining the corona and the solar wind that blows out from it. One experiment will analyze the chemical composition of dust that bounces off the probe's antennae. Another will evaluate the shock waves that rock Sol's plasma, and a third will take a spectral inventory of the plasma's elements.[11] With each new discovery, the Moody Sun Hypothesis will be expanded and refined.

The sudden burst of interest in Sol prompts one to wonder if something more than simple scientific curiosity is going on here, if Destiny might not somehow be taking a hand in our future. Whether or not there is any metaphysical story arc to the Sun–Earth relationship, no doubt can exist that our planet and its sustaining star are deepening their relationship in myriad ways. We are exploring the Sun more vigorously and audaciously than ever before. We are also becoming more vulnerable to the Sun, through the effects of its solar blasts on our satellite system and power grids and through its contribution to global warming. Great demographic shifts toward sunbelt regions of the Northern Hemisphere are making more and more of us Sun worshippers, at least in the lifestyle sense. We are also paying the price for our devotions through the epidemic of skin cancer that overexposure to sunshine has caused. More and more of our power is being drawn directly from the Sun in the form of solar power, a technology that demands an increasingly sophisticated knowledge of sunlight's properties. Every indication is that our future—individually, collectively, and planetarily—will be more and more enmeshed with the Sun.

Not since their dusty, fiery creation have Sol and Gaia been on more intimate terms. Or, as Peaches and Herb sing, "Reunited 'cause we understood."

Conclusion

I have been writing for 25 years now and have come to understand that each book has its own special birthing process. Some cannot wait to leap out of one's head and onto the page, in which case it is necessary to tidy up the mess and fill in the blank spots. Others unfold as though the author were just a transmission vehicle, the airplane that lifts a space shuttle toward orbit and then returns unceremoniously to its hangar. Then there are those oh-so-precious tomes, where each word must be handcrafted, as though squeezed letter by letter out of a pastry bag. *Solar Cataclysm* fit none of these categories; it was an entity unto itself. Writing it was like peeking out over and over again from scary high places, even though doing so always gave me a slippery, prickly vertigo feeling in the soles of my feet. Perhaps this is because the Moody Sun Hypothesis implies a loss of personal sovereignty, arousing the gnawing sense that comes with knowing that yet another entity—be it a deity, Big Brother, or, in this case, the Sun—has control over your life. It's like discovering that you're on a leash 93 million miles long. When is the next yank going to come, and will it break your neck?

It's Copernicus all over again, this time with an edge. One of the reasons his discovery that Earth revolves around the Sun and not vice versa resonates so deeply is that it is so easily personalized. We have all known folks who needed to learn the lesson that the world does not revolve around them. Most of us have gotten taught it a few times ourselves. Now we are told that the star we are eternally stuck orbiting around acts up on occasion. When it does, it can screw up our lives royally, and there's not a damn thing we can do to stop it. The human ego is taking a beating here. But look on the bright side: we have a whole new set of excuses to fall back on now that the Sun is to blame.

"The sunspots made me do it!" In other words, the Sun as caster of shadows. This, of course, is the glass-half-*empty* version of the Moody Sun Hypothesis. The upbeat, can-do approach is every bit as valid as the mopey, the-Sun's-gonna-get-us attitude. One can just as easily take heart in discovering new and dynamic levels of interconnectedness among self, planet, and sustaining star. Sometimes it feels as though we are all moving into an era of heightened Sun-consciousness, mental jukeboxes playing joyous strains of "Let the Sunshine In," "Here Comes the Sun," "You Are the Sunshine of My Life," and "The Sun Will Come Out Tomorrow." Or at least that's how it can seem on the good days.

For the glass-half-*full* interpretation of the Moody Sun Hypothesis—the upbeat interpretation—to be worth its salt, it at least has to enhance our appreciation of daily life and to provide revelatory moments such as the one that enveloped me recently at New York's City Hall, where pedestrians, vehicles, birds, and squirrels swirled around the seat of political power like dancers in a George Balanchine ballet. Each and every movement I saw had, in one way or another, been powered by the Sun. Sunshine was embodied in the produce that people ate in either fresh or processed form, and in the flesh of livestock to which the produce was fed. The fuel burned by cars, buses, and trucks came from oil, the leftover of ancient green plants that eons ago captured the sunlight in order to grow. The odd solar panel drank the juice straight. When it comes right down to it, everything, ultimately, is solar-powered. And everything, ultimately, is subject to changes in that power.

THE NEW EXISTENTIALISM

Each credible threat to civilization comes with its own set of morbid musings. In addition to being insanely murderous, World War II was a wrenching disappointment because it was never supposed to happen in the first place. World War I, "the Big One," was ostensibly "the war to end all wars." In the wake of World War II, midcentury intellectual despair gave rise to existentialism, a bleak but nobly individualistic doctrine in which personal freedom to choose was held to be the utmost value, with the catch that the act of exercising that sacred right irrevocably compromised one's freedom and personal integrity. So to preserve the freedom to choose one had to refrain from choosing. This existential dilemma was presciently expressed in *Nausea* (1938), Jean-Paul Sartre's hilariously unfunny ode to alienation in which the protagonist of the memoir-novel takes a walk, looks around at life, and gets sick to his stomach.

Post-Hiroshima prospects of nuclear Armageddon also gained plenty of intellectual and emotional currency. The "God is dead" movement of the 1960s was built on Nietzsche's belief that mankind had murdered God, and was inflamed by the unprecedented possibilities of megadeath posed by the Cold War arms race. The movement blamed God's death on science and our resulting inability to humble ourselves sufficiently to believe in a transcendent, almighty deity. God simply could not survive our lack of faith. Then came the concept of "nuclear winter." This scientifically iffy but ethically handy proposition advanced by Carl Sagan and others argued that if only one side fired its nukes, even if the other held back entirely, both sides would eventually succumb because all the dust and ash kicked up by the firestorms ignited by the initial assault would blot out the Sun for several years, resulting in plague, famine, and chaos around the world.

That a random belch from the Sun could cripple civilization and trash daily life may well be the next apocalyptic idea to echo like the twang of a broken guitar string through our collective psyche. Solar EMP knocking out the power grid for months or years is the latest, greatest existential threat faced by humankind—far graver than, for example, global warming, if for no other reason than that climate change happens gradually, giving us time to prepare and adjust. You could close this book and go to sleep and then wake up the next morning to find yourself and the rest of society without electricity until next February 29. No weather channel bulletins, no evacuation orders, no FEMA officials stuttering excuses, no state and local authorities protecting each other and blocking off roads. Advance notice, if there is any, will probably reach the public only minutes before the lights go out. That it is within our means, financially and technologically, to shield the power grid from apocalyptic solar EMP, and yet we do not do so, is even more haunting.

It's the perfect setup for a neo-elegiac tradition, wherein the deceased loved one being commemorated is not a person but a civilization. Remember in *Huckleberry Finn* the character of Emmeline Grangerford, the morbid sonneteer who versified eulogies for every corpse in town? She could rhyme darn near any last name and usually beat the undertaker to the deathbed, finished poem in hand, Twain reports. Her masterpiece was "Ode to Stephen Dowling Bots, Dec'd," penned after young Bots fell down a well. Expect eulogies to flow like champagne unless we hurry up and insert those surge suppressors into the grid.

CULTURE, SCHMULTURE

Pondering the solar EMP threat liberated me from contemporary culture at 10:00 A.M. on New Year's Eve day, 2010. While driving home from LAX after dropping off my mother for her flight back to New York City, I listened to the car radio playing yet another "Top 10" end-of-year rundown, this time of the 10 best music albums of 2010. To me, it all sounded overproduced, as though the music were wearing way too much (electronic) makeup. Suddenly, the radio echoed, crackled, and died, much as it would do at the onset of a blackout, I couldn't help thinking. At that moment, drawing distinctions between one pop song and the next seemed profoundly pointless, absurd even, given the possibility that an EMP blast from the Sun could render it all but impossible to make or broadcast such music. Or even to be alive to listen to it. Deck chairs on the *Titanic*.

The death of my car radio put life in perspective. The Mother of All Blackouts, lurking in the background, had somehow freed me from the dictates of today's culture. Freed me from picayune three- and four-star distinctions, from the need to find meaning in the products of the pop entertainment machine, from the pressure to keep up with it all. And that pressure is considerable: I live in L.A., where what's now is passé, and what's next is all. Seriously, how crazily obsessed is our species with its own amusement? Do we care more about Lindsay Lohan scuttlebutt, or, for that matter, the arcane nuances of Wagner's *Ring* cycle, than about our own survival? We have a tendency to become connoisseurs of minutiae while at the same time being lackadaisical in matters of life-and-death importance. This bad habit runs the gamut from the fluff of pop culture to the dry facts of hoary academe. No doubt, arts and entertainment serve as a distraction, as a psychological defense mechanism against being paralyzed by fear. Faced with the prospect of the Mother of All Blackouts, we can pat ourselves on the head and coo sympathetically about the damage that thinking about such awfulness might do to our fragile psyches. Or we can kick the culture habit, at least until we've done everything possible—surge suppressors at a minimum—to ensure the survival of that to which we are so lovingly addicted.

There is no such thing as culture, unplugged. Not anymore— not since movies, television, the Internet, and telecom have taken over. Sure, thoughts and images will continue to be produced in a shorted-out society. Some will be quite artful. Pan-flute pipe dreams of us all singing 'round the campfire might warm the cockles temporarily. Who knows, this could turn out to be the dirge's Golden Age! But the bottom line is that a single, errant outburst from the Sun could fry our precious global electronic culture beyond recognition.

Remember Kris Kristofferson's line in "Bobby McGee," the

song Janis Joplin made famous? "Freedom's just another word for nothin' left to lose."

Okay, so maybe I do still care about really good music. Looks like we all might be set free real soon.

THE SUN AS OUR ENEMY

What if the Super Bowl were hit by the Mother of All Blackouts? Sometimes I'm invited to appear on television documentary programs dealing with solar matters, and if the subject were "The Super Bowl, Unplugged," the director would probably ask me to spell out all the horrors step-by-step. Depending on the show's budget, there might be some impressive, Hollywood-style recreations of a giant solar plasma blob smashing into the stadium, the lights going out, the escalators freezing, 80,000 fans heading for the exits in terror, maybe a poignant close-up of a dazed and confused quarterback standing alone on the field, still slapping the side of his helmet because he can't receive the sideline's radio transmission of the coach calling his next play. All of which would be followed, of course, by the collapse of civilization in the weeks and months ahead, blah, blah, blah.

You think I'm exaggerating? Consider the following passage from a U.S. Army War College report:

Iran, North Korea, China, and Russia have conducted EMP research and, in open source writing, described attacking the United States with EMP. Moreover, Iran has also conducted several missile tests at high altitudes, as if practicing EMP attacks. Yet despite all this evidence that EMP is a clear and present danger, while the USG [United States

government] deserves credit for implementing EMP Commission recommendations to protect U.S. military forces, it deserves condemnation for failing to protect U.S. civilian critical infrastructures from EMP. Given our current state of unpreparedness, within 12 months of an EMP attack or a "great" geomagnetic storm, an estimated two thirds of the U.S. population would perish from starvation and societal collapse.[1]

In the panic of the moment, it wouldn't matter a jot whether the blackout was caused by Nature or by human enemies. But to me, that's the irresistible question, the ultimate projective test. It would be one thing to have the Super Bowl knocked out by a blob of plasma belched by the Sun, quite another if an enemy had set off an EMP device in the airspace above the game. That latter spark would light our emotional fuse. There would be fiery political rhetoric, dramatic news media coverage, massive military mobilization, and a groundswell of support for revenge. We have been attacked by people many times before, so it is easy to imagine being attacked by people again.

What is it about our need for enemies? Has the American consciousness been warped by almost a century of fighting off the Nazis, the Communists, and now Muslim extremists? I ask this question not to minimize the threat from malefactors, which is certainly real enough. September 11th bears stark witness to that. Even the threat of runaway global warming has its villains—us, and our own wastrel ways. My concern is that we have become overly dependent on outrage to mobilize our defenses. This need for hatred or, at the very least, righteous indignation could well cause us to overlook the greatest single threat to our civilization, the turbulent Sun. The Sun is not in any way evil or purposeful, and bears no will—good or ill—toward our species or our planet. In fact, the fluctuations in its energy output have, over the eons, no doubt contributed to the evolutionary outcome we enjoy today. But because we can't get angry at the Sun, can't fear that that yellow ball in the sky actually wants to do us harm, we lack emotional fuel to take the precautions necessary to defend ourselves against its vagaries. By my lights, the hatred we would feel toward an enemy is preferable to the directionless rage of the kind we would experience were we done in by a random, invisible, unintentional blast from space. (Whether our star might somehow serve as the annihilating agent of some higher, sentient power such as God and/or Satan is beyond the scope of human knowledge and therefore of the present inquiry.) So at whom would we shake our fists? How would we vent our rage and grief? Not at merry old Sol. He would never dream of doing such a thing.

Or would he? From Sol's perspective (if he had one), a megablast such as a modern-day recurrence of the Carrington event would not be evil; rather, it would be cleansing in a Darwinian way. By debilitating our current civilization, Sol would, paradoxically, hasten our evolution. Extraordinary dedication and innovation are required to rebuild stronger, newer, and better societies after long-standing civilizations have been destroyed. For example, earthquakes level cities; then cities rebound with superior, earthquake-proof structures. Plagues decimate populations; those who survive create offspring with strengthened immune systems so that a greater share of the populace will survive the next plague. In short, Sol might well see catastrophe as the organizing principle of evolution, and himself as the principal provider of catastrophe. Any surge suppressor measures to prevent global cataclysm would, from his perspective, impede humankind's ascent. Of course, most of us are in no particular hurry to climb the evolutionary ladder, thank you. We kind of like the rung we're on. To us, it's far more important to preserve what we have and to spare innocent people and creatures from suffering and death.

Maybe it is simply a coincidence that the Sun is being examined up, down, and sideways by IHY spacecraft these days, just as the great ongoing solar storm of the past century and a half reaches its next climax, and just as our technological infrastructure enters the red zone of its vulnerability to solar blasts. Maybe not. This book makes no teleological assumptions, observing only that the timing of all this IHY space research is very fortunate. Suspiciously fortunate. There is something providential in the way things are working out. After umpteen millennia, we finally develop the technological ability to examine the Sun close up and firsthand, to see what really makes it tick and what really makes it dangerous. We gain this knowledge, knock wood, just in the nick of time, since never before in those umpteen millennia have we been nearly so vulnerable to the Sun's blasts. Again technology is the key variable, though in this second instance because of its mortal susceptibility to solar EMP. Is it just an accident that these two timelines—our vulnerability to the Sun and our knowledge of it—are playing out in unison?

Are we playing a game of chicken with ourselves? Or does the synchronous progression of capability and susceptibility somehow jibe with the natural order of things? Wouldn't it be lovely if the doctrine to emerge from this latest threat to our existence were that the universe provides vital information on a need-to-know basis, even if we don't consciously know that we need to know it? That just as the Sun threatens our world, we are given the technical knowledge necessary to blunt that threat and survive the nip-and-tuck survival race. How life-affirming that would be!

IMPLORING THE SUN

What of the proposition that we can influence the Sun?

Science rarely intersects with whimsy, the former being dismissive of the latter and the latter, ooh la la, aching to dab a touch of glitter on the former's furrowed brow. But the very idea that the Sun is mindlessly, haphazardly in charge of so much more than we ever realized demands some kind of nonlinear response, doesn't it? Even if it is physically impossible to calm, charm, or otherwise influence the Sun's behavior, we still have a sacred, bounden duty to try anyway. I have no doubt that worship will help. Illogical? You betcha! Then again, not entirely. Prayer is a superb way to focus one's sensitivity, elevating one's ability to respond to whatever our sustaining star might cast our way. For example, I give thanks to a variety of deities and transcendent entities, including the universe, to which I say, "You are the greatest thing; you are everything; you are the only thing. May I always be in harmony with your intentions." Do I believe that the universe hears my prayer? No, the universe, which is simply the totality of matter and energy, is not conscious. Do I nonetheless believe that intoning these sentences will help spare me from harm and/or bring me benefit? Absolutely. And I have the strong feeling that imploring the universe, and its local overlord, the Sun, will advance your cause too.

What might we ask of the Sun? While it would be presumptuous to think that we have the ability to determine optimal levels of solar behavior for ourselves or our planet, much less the star itself, it is fair to say that we prosper most when the Sun does not deviate too far from its baseline level of activity. Spikes and troughs should not be extreme in amplitude or duration. No more frigid minima like the Maunder Minimum, please, and no more megamaxima of the type that melted the most recent Ice Age and that would roast us extra-crispy today. Good, now where do we file these requests? Communing with the Sun, or with anything else for that matter, is not a process of negotiation or even of asking nicely. Rather, it is a unification in which all partners combine, sometimes complementarily, other times in parallel, to form a more vibrant whole. To that end, I would like to reintroduce Circles of Mud, a simple ritual first proposed in my recent book *Aftermath*, wherein one person draws three circles of mud around another person's neck in solemn observation of our sacred connection to Sun and Earth.

The first circle of mud symbolizes a wedding ring, for the marriage of head to heart, Sun to Earth, birth to death.

The second circle of mud symbolizes the beheading of all thoughts that separate us from the goodness of life.

The third circle of mud symbolizes eternity and the joy that comes with knowing that end follows beginning and beginning follows end, in an eternal orbit.

Family members, friends, neighbors, strangers—anyone may anoint and anyone may be anointed. Mud should be left on the skin until it dries, preferably in the sunlight. Participating in this ceremony does not affect the celebrant's religious status, nor does it commit him or her to any set of beliefs other than the hope that Sol, Gaia, and their children will get along harmoniously for centuries to come.

WHILE WE'RE AT IT . . .

While we're at it, let's sacrifice a few surge suppressors as an offering to Sol. Bottom line, there is no chance that the North American power grid will have been protected from EMP, solar or otherwise, before the 2012–2013 climax. The best we can hope for is to luck out and dodge any rogue storms that might come along, and during the ensuing decade make sure to get ourselves ready for the subsequent solar tantrum, due in 2023–2024. That's a reasonable goal. Fingers crossed, or interlaced in prayer, that we get the job done in time.

NOTES

INTRODUCTION

1. Ilya G. Usoskin, "A History of Solar Activity over Millennia," *Living Reviews in Solar Physics* 5, no. 3 (2008), http://www.livingreviews.org/lrsp-2008-3, Figure 14.
2. Brian Fagan, *The Little Ice Age: How Climate Made History, 1300–1850* (New York: Basic Books, 2000), 107.
3. Jared Diamond, *Collapse: How Societies Choose to Fail or Succeed* (New York: Viking, 2005).
4. National Academy of Sciences/NASA, *Severe Space Weather Events: Understanding Societal and Economic Impacts* (Washington, D.C.: National Academies Press, December 2008), 77.

SPECIAL NOTE: THE MOODY SUN HYPOTHESIS

1. Margaret Cheney, *Tesla: Man Out of Time* (New York: Dell, 1981), 55–56.
2. Tony Phillips, "Global Eruption Rocks the Sun," *NASA Science News* (December 13, 2010), http://science.nasa.gov/science-news/science-at-nasa/2010/13dec_globaleruption/.
3. Thomas Kuhn, *The Structure of Scientific Revolutions*, 2nd ed., enlarged (Chicago: Univ. of Chicago Press, 1970), 113.

CHAPTER 1: SOL HAS LOVED GAIA FOR ALMOST FIVE BILLION YEARS

1. Carl Sagan and George Mullen, "Earth and Mars: Evolution of Atmospheres and Surface Temperatures," *Science* 177 (1972): 52–56.
2. Emanuel Swedenborg, *The New Jerusalem and Its Heavenly Doctrine* (1758), translated from the Latin in 1911 (London, England: The Swedenborg Society), 261.

3. James Lovelock, *The Ages of Gaia: A Biography of Our Living Earth* (New York: W.W. Norton and Company, 1988), 17.
4. Lynn Margulis and Dorion Sagan, *Microcosmos: Four Billion Years of Microbial Evolution* (New York: Summit Books, 1986), 55–56.
5. Margulis and Sagan, *Microcosmos*, 29.

CHAPTER 2: SUNSPOTS MELTED THE LAST ICE AGE
1. E. H. Gombrich, *A Little History of the World* (New Haven, CT: Yale Univ. Press, 2005), 8.
2. John A. Garraty and Peter Gay, eds., *The Columbia History of the World* (New York: Harper and Row, 1972).
3. Philippe Aries and Georges Duby, eds., *A History of Private Life: From Pagan Rome to Byzantium* (Cambridge, MA: Belknap/Harvard University Press, 1987), 183.
4. Brian Fagan, *The Great Warming: Climate Change and the Rise and Fall of Civilizations* (New York: Bloomsbury Press, 2008), 11.
5. Garraty and Gay, *Columbia History of the World*, 47.
6. Garraty and Gay, *Columbia History of the World*, 46.

CHAPTER 3: THE LONG, PAINFUL HISTORY OF SUNSPOT DENIAL
1. Aristotle, *On the Heavens*, 1.10–12.
2. Wikipedia: Encyclopaedia Britannica (1777), Dr. Long's copy of Cassini, http://en.wikipedia.org/wiki/File:Cassini_apparent.jpg.
3. Garraty and Gay, *Columbia History of the World*, 516.
4. Sheila Rabin, "Nicolaus Copernicus," *Stanford Encyclopedia of Philosophy*, http://plato.stanford.edu/entries/copernicus/.
5. Nicolaus Copernicus, *On the Revolutions of the Celestial Spheres* (1543, 1.10).
6. Kuhn, *The Structure of Scientific Revolutions*, 116–117.
7. Galileo Galilei, "Third Letter on Sunspots, from Galileo Galilei to Mark Welser," *Discoveries and Opinions of Galileo*, trans. Stillman Drake (Garden City, NY: Doubleday Anchor Books, 1957), 122–44.
8. MMVSEC, 2.

CHAPTER 4: FROM THE MEDIEVAL WARM PERIOD TO THE LITTLE ICE AGE
1. Brian Fagan, *The Little Ice Age: How Climate Made History, 1300–1850* (New York: Basic Books/Perseus, 2000), 13–15.
2. Fagan, *The Little Ice Age*, 13–15.
3. Willie Soon and Sallie Baliunas, *Lessons and Limits of Climate History: Was the Twentieth Century Climate Unusual?* (Washington, D.C.: George C. Marshall Institute, 2003), 16.
4. Stephen McIntyre and Ross McKitrick, "Corrections to the Mann et al. (1998) Proxy Data Base and Northern Hemispheric Average Temperature Series," *Energy & Environment* 14, no. 6 (2003): 751–71.
5. Richard A. Muller, "Global Warming Bombshell," *Technology Review* (October 15, 2004), http://www.technologyreview.com/energy/13830.
6. Fagan, *The Great Warming*, 14–15.

7. Simon Ings, A *Natural History of Seeing: The Art and Science of Vision* (New York: Norton, 2007), 27.
8. Fagan, *The Great Warming*, 7, 11.
9. Stuart Clark, *The Sun Kings: The Unexpected Tragedy of Richard Carrington and the Tale of How Modern Astronomy Began* (Princeton, NJ: Princeton Univ. Press, 2007), 182.
10. Sultan Hameed and Gaofa Gong, "Prolonged Drought in Northern China During the Maunder Minimum and Its Relation to Peasant Rebellions and Fall of the Ming Dynasty," conference presentation, http://lasp.colorado.edu/fri_am/Hameed_China_Drought.pdf.
11. Diamond, *Collapse*, chap. 2.
12. Bradd Shore, "Rapanui: Tradition and Survival on Easter Island. Grant McCall," *American Anthropologist* 84 (October 28, 2009): 717–18.
13. "Study of Dust in Ice Cores Shows Volcanic Eruptions Interfere with the Effect of Sunspots on Global Climate," University at Buffalo news release (June 11, 2002), http://www.buffalo.edu/news/fast-execute.cgi/article-page.html?article=57350009.
14. Ram, "Study of Dust in Ice Cores," x.
15. National Center for Atmospheric Research/University of Colorado–Boulder, "Was the Little Ice Age Triggered by Massive Volcanic Eruptions?" *Science Daily* (January 30, 2012), http://www.sciencedaily.com/releases/2012/01/120130131509.htm.

CHAPTER 5: WHEN THE SUN FELL ASLEEP

1. John A. Eddy, "The Maunder Minimum," *Science* (June 18, 1976), 1189–1202, http://www.sciencemag.org/content/192/4245/1189.citation.
2. Ellen McClure, *Sunspots and the Sun King: Sovereignty and Mediation in Seventeenth Century France* (Urbana and Chicago: Univ. of Illinois Press, 2006), 1, 2.
3. McClure, *Sunspots and the Sun King*, 61.
4. Quoted in McClure, *Sunspots and the Sun King*, 64.
5. McClure, *Sunspots and the Sun King*, 66.

CHAPTER 6: SUNSPOTS ARE MAKING THE WORLD WARMER

1. Tony Phillips, "Solar Dynamics Observatory: The 'Variable Sun' Mission," *Science News* (February 5, 2010), http://science.nasa.gov/science-news/science-at-nasa/2010/05feb_sdo/.
2. Amy Bingham, "Al Gore Calls BS on Climate Change Doubters," *ABC News* (August 9, 2011), http://abcnews.go.com/blogs/politics/2011/08/al-gore-calls-bs-on-climate-change-doubters/.
3. George Bush, remarks at the Univ. of Michigan Commencement Ceremony in Ann Arbor (May 4, 1991), http://www.presidency.ucsb.edu/ws/index.php?pid=19546&st=home+ownership&st1=.
4. Phillips, "Solar Dynamics Observatory," http://science.nasa.gov/science-news/science-at-nasa/2010/05feb_sdo/.

5. Gary Rottman and Robert Calahan, "SORCE: Solar Radiation and Climate Experiment," Laboratory for Atmospheric and Space Physics (LASP), Univ. of Colorado, and NASA Goddard Space Flight Center (2004): 2.

6. Willie Soon, "Solar Variability and Climate Change," Marshall Institute (January 10, 2000), http://www.marshall.org/article.php?id=91.

7. "Ron Johnson: DSCC Attack," *Freedom Eden* (August 25, 1010), http://freedomeden.blogspot.com/2010/08/ron-johnson-dscc-attack.html.

8. Kenneth R. Lang, *Cambridge Encyclopedia of the Sun* (Cambridge, UK: Cambridge Univ. Press, 2001), 180.

9. "Everest Could Soon Become Impossible to Climb Because of Global Warming, Says Top Sherpa," *Mail Online*, http://www.dailymail.co.uk/sciencetech/article-2107013/.

10. Seth Borenstein, "Skeptic Finds He Now Agrees Global Warming Is Real," *Boston.com* (October 30, 2011), http://articles.boston.com/2011-10-30/news/30339465_1/.

11. Richard Michael Pasichnyk, "The Unity of the Sun, Earth, and Moon," *The Living Cosmos*, http://www.livingcosmos.com/unity.htm.

12. Pasichnyk, "The Unity of the Sun, Earth, and Moon," 5.

13. Vladimir I. Vernadsky, *The Biosphere* (New York: Springer-Verlag, 1997), 44.

14. Douglas V. Hoyt and Kenneth H. Schatten, *The Role of the Sun in Global Warming* (New York: Oxford University Press, 1997), 3–4.

15. Lang, *Cambridge Encyclopedia of the Sun*, 182–83.

CHAPTER 7: THE GLOBAL COOLING SCANDAL

1. "British Climatic Research Unit's Emails Hacked," *Wikinews* (November 19, 2009), http://en.wikinews.org/wiki/British_Climatic_Research_Unit%27s_emails_hacked.

2. Clive Crook, "Climategate Corrections and Revisions 2," http://www.theatlantic.com/politics/archive/2010/09/climategate-corrections-and-revisions-2/63071/.

3. Crook, "Climategate Corrections and Revisions 2."

4. "The Role of Sunspots and Solar Winds in Climate Change," *Scientific American* (July 22, 2009), http://www.scientificamerican.com/article.cfm?id=sun-spots-and-climate-change.

5. Mao Tse-Tung, www.abovetopsecret.com/forum/threads514758/pg1.

6. Tony Phillips, "Researchers Crack the Mystery of the Missing Sunspots," *NASA Science News* (March 2, 2011), http://science.nasa.gov/science-news/science-at-nasa/2011/02mar_spotlesssun/.

7. Tony Phillips, "A Puzzling Collapse of Earth's Upper Atmosphere," *NASA Science News* (July 15, 2010), http://science.nasa.gov/science-news/science-at-nasa/2010/15jul_thermosphere/.

8. Tony Phillips, "Cosmic Rays Hit Space Age High," *NASA Science News* (September 20, 2009), http://science.nasa.gov/science-news/science-at-nasa/2009/20sep_cosmicrays.

9. Phillips, "Cosmic Rays Hit Space Age High."

CHAPTER 8: TWENTY-FIRST-CENTURY SUN WORSHIP

1. *What You Need to Know About™ Melanoma and Other Skin Cancers*, National Cancer Institute booklet, http://www.cancer.gov/cancertopics/wyntk/skin.
2. "Skin Cancer Prevention and Early Detection," American Cancer Society, http://www.cancer.org/Cancer/CancerCauses/SunandUVExposure/.
3. *Ultraviolet Radiation and Melanoma: With a Special Focus on Assessing the Risks of Stratospheric Ozone Depletion*, United States Environmental Protection Agency (December 1987).
4. Andreas Schweizer, *The Sungod's Journey Through the Netherworld: Reading the Ancient Egyptian Amduat* (Ithaca, NY: Cornell Univ. Press, 2010).
5. Robert Cox, *The Pillar of Celestial Fire: The Lost Science of the Ancient Seers Rediscovered* (Fairfield, IA: Sunstar Publishing, 1997), 71.

CHAPTER 9: THE SUN ALSO HEALS YOU

1. "Sunlight and Sungazing," sunlight.orgfree.com/sungazing.htm.
2. Ings, *A Natural History of Seeing*, 29–30.
3. Ings, *A Natural History of Seeing*, 28.
4. Leonard I. Grossweiner, *The Science of Phototherapy: An Introduction* (Dordrecht, Netherlands: Springer-Verlag, 2010), 3.
5. Grossweiner, *The Science of Phototherapy*, ix.
6. Cristina Luigi, "Light Therapy, Circa 1939," *Scientist: Magazine of the Life Sciences* (February 1, 2011), http://www.the-scientist.com/article/display/57954/.
7. *Nobel Lectures, Physiology or Medicine 1901–1921* (Amsterdam: Elsevier Publishing Company, 1967), http:www.nobelprize.org/nobel_prizes/medicine/laureates/1903/finsen-bio.html.
8. Grossweiner, *The Science of Phototherapy*, 3.
9. "Dermatologists' Perspective on Myths and Facts About Vitamin D and Sun Exposure," *Health Gazette* (May 11, 2006), http://the-health-gazette.com/469/dermatologists-perspective-on-myths-and-facts-about-vitamin-d-and-sun-exposure/.
10. "Dietary Supplement Fact Sheet: Vitamin D," National Institutes of Health, http://ods.od.nih.gov/factsheets/VitaminD-QuickFacts/.
11. "Vitamin D," Mayo Clinic, http://www.mayoclinic.com/health/vitamin-d/NS_patient-vitamind/.
12. Institute of Medicine, National Academy of Sciences, *Dietary Reference Intakes for Calcium and Vitamin D* (November 30, 2010), http://www.iom.edu/Reports/2010/Dietary-Reference-Intakes-for-Calcium-and-Vitamin-D.aspx.
13. Rebecca Smith, "Vitamin D Can Aid Fertility," *The Telegraph* (November 11, 2008), http://www.telegraph.co.uk/health/women_shealth/3434420/Vitamin-D-can-aid-fertility.html.
14. Smith, "Vitamin D Can Aid Fertility."
15. Rachel Champeau, "Scientists Find Vitamin D Is Crucial in Human Immune Response to Tuberculosis," *UCLA Newsroom* (October 12, 2011), http://newsroom.ucla.edu/portal/ucla/scientists-find-vitamin-d-crucial-216881.aspx.

16. "Prostrate Disorders Special Report: Vitamin D and Prostrate Cancer," *Johns Hopkins Medicine Health Alerts* (July 23, 2009), http://www.johnshopkins healthalerts.com/reports/prostate_disorders/3115-1.html.

CHAPTER 10: SUNSPOTS AND YOUR BRAIN

1. Anna Krivelyova and Cesare Robotti, "Playing the Field: Geomagnetic Storms and the Stock Market" (working paper 2003-5b, Federal Reserve Bank of Atlanta, October 2003), http://www.frbatlanta.org/filelegacydocs/wp0305b.pdf, 1.
2. Marcia Barinaga, "Giving Personal Magnetism a Whole New Meaning," *Science* 256 (May 15, 1992): 967.
3. Neil Cherry, "Human Intelligence: The Brain, an Electromagnetic System Synchronised by the Schumann Resonance Signal," *Medical Hypotheses* 60 (June 2003): 843–44.
4. Catherine Brahic, "Does the Earth's Magnetic Field Cause Suicides?" *New Scientist* (April 24, 2008), http://www.newscientist.com/article/dn13769-does-the-earths-magnetic-field-cause-suicides.html.
5. Steele Hill and Michael Carlowicz, *The Sun* (New York: Abrams, 2006): 26.
6. Brahic, "Does the Earth's Magnetic Field Cause Suicides?"
7. Amir Raz, "Could Certain Frequencies of Electromagnetic Waves or Radiation Interfere with Brain Function?" *Scientific American* (April 24, 2006): 14.
8. Brahic, "Does the Earth's Magnetic Field Cause Suicides?"
9. Krivelyova and Robotti, "Playing the Field: Geomagnetic Storms and the Stock Market," 1.
10. "Stock Indexes Have Best January Since 1997," *GazetteNet.com* (February 1, 2012), http://www.gazettenet.com/2012/02/01/stock-indexes-have-best-january-since-1997.
11. Adrian Chen, "Does Sad Obama Have Seasonal Affective Disorder?" (March 16, 2010), http://gawker.com/5494258/does-sad-obama-have-seasonal-affective-disorder.
12. Alexander V. Trofimov, personal interview.
13. "The Water in You," USGS Water Science for Schools, http://ga.water.usgs.gov/edu/propertyyou.html.

CHAPTER 11: THE SUN WILL SOON SHORT OUT THE ELECTRICAL POWER GRID

1. Georges Lakhovsky, "Influence of Cosmic Waves on the Oscillation of Living Cells," paper presented to the Academie des Sciences by Professor d'Arsonval (March 28, 1927), http://multiplewaveoscillator.com/cosmicinfluenceMWO1.html.
2. Roland Barthes, "Wine and Milk," *Mythologies*, trans. Annette Lavers (New York: Hill and Wang, 1972), 58.
3. Hoyt and Schatten, *The Role of the Sun in Climate Change*, 4.
4. William Stanley Jevons, *Investigations in Currency and Finance* (London: Macmillan, 1909).

5. North American Electric Reliability Corporation and the U.S. Department of Energy, *High-Impact, Low-Frequency Event Risks to the North American Bulk Power System* (June 2010), http://www.nerc.com/files/HILF.pdf.

6. Electromagnetic Pulse Commission (of the U.S. government), *Critical National Infrastructures Report* (July 2008), http://empcommission.org/docs/A2473-EMP_Commission-7MB.pdf.

CHAPTER 12: A SIMPLE WAY TO PROTECT OUR FUTURE

1. John Kappenman, personal interview.

2. Lawrence E. Joseph, "The Sun Also Surprises," *New York Times* (August 15, 2010), http://www.nytimes.com/2010/08/16/opinion/16joseph.html.

3. Thomas Popik, prepared statement, Foundation for Resilient Societies (February 8, 2011): 6.

4. Frequently Asked Questions, EMPact America, http://www.empactamerica.org/faq.php.

5. Peter Asmus, "Why Microgrids Are Inevitable," *Distributed Energy: The Journal of Energy Efficiency and Reliability* (September–October 2011), http://www.distributedenergy.com/DE/Articles/Why_Microgrids_Are_Inevitable_15471.aspx. The subsequent quote is also from this source.

6. Joe Davila, quoted in Tony Phillips, "The International Space Weather Initiative," *NASA Science News* (November 8, 2010), http://science.nasa.gov/science-news/science-at-nasa/2010/08nov_iswi/.

7. Tony Phillips, "The International Space Weather Initiative," *NASA Science News* (November 8, 2010), http://science.nasa.gov/science-news/science-at-nasa/2010/08nov_iswi/.

CHAPTER 13: A HUNDRED NUCLEAR MELTDOWNS COMING OUR WAY

1. Thomas Popik, Petition for Rulemaking before the United States Nuclear Regulatory Commission by the Foundation for Resilient Societies, February 6, 2011.

2. John Kappenman, "Geomagnetic Storms and Their Impacts on the U.S. Power Grid," report prepared for Oak Ridge National Laboratory (January 2010): 1–17, http://www.fas.org/irp/eprint/geomag.pdf.

CHAPTER 14: THE SUN WILL SEND US SECRET WARNINGS

1. Dan Stober, "The Strange Case of Solar Flares and Radioactive Elements," *Stanford University News* (August 23, 2010), http://news.stanford.edu/news/2010/august/sun-082310.html. The subsequent five quotes are also from this source.

CHAPTER 15: THREE LOOMING THREATS AND ONE HAPPY ENDING

1. David Sibeck, "Giant Breach in Earth's Magnetic Field Discovered" (December 16, 2008), http://science.nasa.gov/headlines/y2008/16dec_giantbreach.htm?friend.

2. Robert Sanders, "Gamma-Ray Flash Came from Star Being Eaten by Massive Black Hole," *Science Daily* (June 16, 2011), http://www.sciencedaily.com/releases/2011/06/110616142709.htm.

3. Merav Opher, "A Strong, Highly-Tilted Interstellar Magnetic Field Near the Solar System," *Nature* 462 (December 24, 2009), http://www.nature.com/nature/journal/v462/n7276/full/nature08567.html. The subsequent three quotes are also from this source.
4. Alexey N. Dmitriev, "Planetophysical State of the Earth and Life," trans. A. N. Dmitriev, Andrew Tetenov, and Earl L. Crockett, English presentation sponsored by The Millennium Group (January 8, 1998), http://tmgnow.com/repository/global/planetophysical.html.
5. M. Davis, P. Hut, and R. Muller, "Extinction of Species by Periodic Comet Showers," *Nature* 308 (1984): 715–717.
6. James W. Kirchner and Anne Weil, "Fossils Make Waves," *Nature* 434 (March 10, 2005): 147–148.
7. Walter Cruttenden, "Precession of the Equinox," Binary Research Institute, http://www.binaryresearchinstitute.org/bri/research/introduction/precession.shtml.
8. Cal Fussman, "The Man Who Finds Planets," *Discover Magazine* (May 27, 2006), http://discovermagazine.com/2006/may/cover/.
9. Arthur Eddington, *The Internal Constitution of the Stars* (Cambridge, U.K.: Cambridge Univ. Press, 1988).
10. NASA, "NASA Selects Science Investigations for Solar Probe Plus" (September 2, 2010), http://www.nasa.gov/topics/solarsystem/sunearthsystem/main/solarprobeplus.html.
11. NASA, "NASA Selects Science Investigations for Solar Probe Plus."

CONCLUSION

1. Kevin Cogan, *In the Dark: Military Planning for a Catastrophic Critical Infrastructure Event* (Carlisle, PA: U.S. Army War College, May 2011), 18.

SUGGESTED READING

Barthes, Roland. *Mythologies*. Translated by Annette Lavers. New York: Hill and Wang, 1972.

Brody, Judit. *The Enigma of Sunspots: A Story of Discovery and Scientific Revolution*. Edinburgh, UK: Floris Books, 2002.

Cheney, Margaret. *Tesla: Man Out of Time*. New York: Dell, 1981.

Clark, Stuart. *The Sun Kings: The Unexpected Tragedy of Richard Carrington and the Tale of How Modern Astronomy Began*. Princeton, NJ: Princeton Univ. Press, 2007.

Cox, Robert. *The Pillar of Celestial Fire: The Lost Science of the Ancient Seers Rediscovered*. Fairfield, IA: Sunstar Publishing, 1997.

Durant, Will. *The Story of Philosophy: The Lives and Opinions of the Greatest Philosophers*. New York: Simon and Schuster, 1961.

Fagan, Brian. *The Great Warming: Climate Change and the Rise and Fall of Civilizations*. New York: Bloomsbury Press, 2008.

——. *The Little Ice Age: How Climate Made History, 1300–1850*. New York: Basic Books, 2000.

Garraty, John A., and Peter Gay, eds. *The Columbia History of the World*. New York: Harper and Row, 1972.

Golub, Leon, and Jay M. Pasachoff. *Nearest Star: The Surprising Science of Our Sun*. Cambridge, MA: Harvard University Press, 2002.

Gombrich, E. H. *A Little History of the World*. New Haven, CT: Yale Univ. Press, 2005.

Grossweiner, Leonard I. *The Science of Phototherapy: An Introduction*. Dordrecht, Netherlands: Springer-Verlag, 2010.

Harvey, Sir Paul. *The Oxford Companion to English Literature.* Oxford, UK: Oxford Univ. Press, 1967.

Hill, Steel, and Michael Carlowicz. *The Sun.* New York: Abrams, 2006.

Ings, Simon. *A Natural History of Seeing: The Art and Science of Vision.* New York: Norton, 2007.

Joseph, Lawrence E. *Aftermath: A Guide to Preparing for and Surviving Apocalypse 2012.* New York: Broadway Books, 2010.

———. *Apocalypse 2012: A Scientific Investigation into Civilization's End.* New York: Broadway Books, 2007.

———. *Common Sense: Why It's No Longer Common.* New York: Addison-Wesley, 1994.

———. *Gaia: The Growth of an Idea.* New York: St. Martin's Press, 1990.

Kuhn, Thomas S. *The Structure of Scientific Revolutions.* 2nd ed., enlarged. Chicago: Univ. of Chicago Press, 1970.

Lang, Kenneth R. *The Cambridge Encyclopedia of the Sun.* Cambridge, UK: Cambridge Univ. Press, 2001.

McClure, Ellen M. *Sunspots and the Sun King: Sovereignty and Mediation in Seventeenth Century France.* Urbana and Chicago: Univ. of Illinois Press, 2006.

National Research Council of the National Academies. *Severe Space Weather Events: Understanding Societal and Economic Impacts: A Workshop Report.* Washington, D.C.: National Academies Press, 2008.

Pasichnyk, Richard Michael. *In Defense of Nature: The History Nobody Told You About.* Lincoln, NE: Writer's Club Press / iUniverse, 2003.

———. *The Vital Vastness.* Vol. 1, *Our Living Earth.* Lincoln, NE: Writer's Showcase / iUniverse, 2002.

———. *The Vital Vastness.* Vol. 2, *The Living Cosmos.* Lincoln, NE: Writer's Showcase / iUniverse, 2002.

Schweizer, Andreas. *The Sungod's Journey Through the Netherworld: Reading the Ancient Amduat.* Ithaca, NY: Cornell Univ. Press, 2010.

Vernadsky, Vladimir I. *The Biosphere.* Complete annotated ed. New York: Springer-Verlag, 1997.

Veyne, Paul, ed. *A History of Private Life: From Pagan Rome to Byzantium.* Cambridge, MA: Belknap / Harvard Univ. Press, 1987.

Whittock, Martyn. *A Brief History of Life in the Middle Ages: Scenes from the Town and Countryside of Medieval England.* Philadelphia, PA: Running Press, 2009.

Wilson, Edward O. *Consilience: The Unity of Knowledge.* New York: Knopf, 1998.

Wolf, John B. *Louis XIV.* New York: Norton, 1968.

ACKNOWLEDGMENTS

Basically, I wouldn't have looked into any of this were it not for John Kappenman, a crusading electrical engineer from Duluth, Minnesota, who has dedicated his career to protecting the power grid that sustains our civilization.

The Moody Sun Hypothesis propounded in this book comes directly from my attempt to emulate James Lovelock, the genius who invented the Gaia Hypothesis, a profoundly innovative way to think about life on Earth.

Jeanette Perez, my editor, has been gracious, wise, and helpful, even during this, the busiest time of her life thus far.

Andrew Stuart, literary agent extraordinaire, redefines stalwart and is owed a great deal.

Thank you Mom, Phoenix, Erica, and Elia for helping me keep my eyes on the prize during this blind-man's bluff of a year.

INDEX